# Smart Grid Technology

Growing concern for climate change and energy conservation is making people focus on digitalized and green technologies for modern power delivery systems. The smart grid is conceptualized as the integration of bi-directional communication networks over the existing electricity networks. Cloud computing and data management approaches are suitable additions to smart grids, to support real-time energy management in a cost-effective, reliable and secure manner.

This book offers comprehensive coverage on smart grid technologies, their concepts and underlying principles. It integrates the tools and techniques of cloud computing and data management for application in smart grids. Different cloud computing and data management approaches are explained highlighting energy management, information management, and security in the smart grid. The concepts of plug-in hybrid electric vehicle and virtual energy storage are explained in separate chapters. The text covers recent trends in cloud computing and data analytics in the field of smart grid. A glossary of important technical terms is provided for the benefit of the readers.

**Sudip Misra** is a Professor in the Department of Computer Science and Engineering at the Indian Institute of Technology Kharagpur. He has co-authored over 300 scholarly research papers. He has also published 9 books in the areas of wireless networks. He is a Fellow of the National Academy of Science India (NASI). Sudip is an associate editor of the IEEE Transactions on Mobile Computing, IEEE Transactions on Vehicular Technology, and IEEE Systems Journal. His current research interests include wireless ad-hoc and sensor networks, internet of things (IoT), computer networks, and smart grid communications. More information about him is available at http://cse.iitkgp.ac.in/~smisra/.

**Samaresh Bera** works as a Senior Researcher at the Smart Wireless Applications and Networking (SWAN) Lab of the Department of Computer Science and Engineering, Indian Institute of Technology Kharagpur. He has published several papers in peer reviewed journals on smart grid, parallel and distributed systems. His current research interests include software-defined networking, smart grid communications and networking, and internet of things (IoT). More information about him is available at http://cse.iitkgp.ac.in/~sambera/.

# Smart Grid Technology

*A Cloud Computing and Data Management Approach*

**Sudip Misra**
**Samaresh Bera**

# CAMBRIDGE
## UNIVERSITY PRESS

University Printing House, Cambridge CB2 8BS, United Kingdom

One Liberty Plaza, 20th Floor, New York, NY 10006, USA

477 Williamstown Road, Port Melbourne, VIC 3207, Australia

314 to 321, 3rd Floor, Plot No.3, Splendor Forum, Jasola District Centre, New Delhi 110025, India

79 Anson Road, #06–04/06, Singapore 079906

Cambridge University Press is part of the University of Cambridge.

It furthers the University's mission by disseminating knowledge in the pursuit of education, learning and research at the highest international levels of excellence.

www.cambridge.org
Information on this title: www.cambridge.org/9781108475204

© Sudip Misra and Samaresh Bera 2018

This publication is in copyright. Subject to statutory exception and to the provisions of relevant collective licensing agreements, no reproduction of any part may take place without the written permission of Cambridge University Press.

First published 2018

Printed in India

*A catalogue record for this publication is available from the British Library*

ISBN: 978-1-108-47520-4 Hardback

Additional resources for this publication at www.cambridge.org/9781108475204

Cambridge University Press has no responsibility for the persistence or accuracy of URLs for external or third-party internet websites referred to in this publication, and does not guarantee that any content on such websites is, or will remain, accurate or appropriate.

*Sudip dedicates this book to his grandparents and grandparents-in-law*
*Samaresh dedicates this book to his newborn daughter Samapika*

# Contents

*Figures* xv
*Tables* xix
*Foreword* xxi
*Preface* xxv

## Part I
## Introduction

**1. Introduction to Smart Grid**   3
  1.1 Smart Grid Framework and Communication Model   3
  1.2 Smart Grid Vision   6
  1.3 Requirements of a Smart Grid   8
    1.3.1 Energy management   8
    1.3.2 Need to support multiple devices   9
    1.3.3 Information management   9
    1.3.4 Layered architecture   10
    1.3.5 Security   10
  1.4 Components of the Smart Grid   11
    1.4.1 Bi-directional communication   11
    1.4.2 Smart meter   12
    1.4.3 Micro-grid   13
    1.4.4 Plug-in hybrid electric vehicles   14
  1.5 Smart Grid Interoperability   14

1.6　Summary　15
　　　　References　17

## 2. Introduction to Cloud Computing　18
　　2.1　Allied Computing Models　20
　　　　2.1.1　Mainframes　20
　　　　2.1.2　Client–server architecture　23
　　　　2.1.3　Cluster computing　25
　　　　2.1.4　Grid computing　26
　　　　2.1.5　Service oriented architecture (SOA)　27
　　　　2.1.6　Utility computing　28
　　　　2.1.7　Pay-per-use model　28
　　2.2　Virtualization　28
　　2.3　Hypervisor　29
　　2.4　Types of Services　31
　　　　2.4.1　Infrastructure as a service　31
　　　　2.4.2　Platform as a service　31
　　　　2.4.3　Software as a service　32
　　2.5　Types of Deployment　33
　　　　2.5.1　Public cloud　33
　　　　2.5.2　Private cloud　33
　　　　2.5.3　Hybrid cloud　34
　　　　2.5.4　Community cloud　34
　　2.6　Advantages of Cloud Computing　34
　　　　2.6.1　Elastic nature　34
　　　　2.6.2　Shared architecture　35
　　　　2.6.3　Metering architecture　35
　　　　2.6.4　Supports existing internet services　35
　　2.7　Architecture of Cloud Computing　35
　　2.8　Summary　36
　　　　References　37

## 3. Introduction to Big Data Analytics　38
　　3.1　Attributes of Big Data　39
　　　　3.1.1　Volume of data　39
　　　　3.1.2　Velocity of data　40

|   |   |
|---|---|
| 3.1.3 Variety of data | 41 |
| 3.2 Overview of Big Data Analytics | 41 |
| 3.3 Benefits of Big Data Analytics | 44 |
| 3.4 Big Data Analytics for Smart Grid | 45 |
| 3.5 Big Data Analytics Tools | 46 |
| 3.6 Summary | 47 |
| References | 47 |

## 4. Fundamental Mathematical Prerequisites — 49

|   |   |
|---|---|
| 4.1 Linear Programming | 49 |
| 4.2 Integer Linear Programming | 50 |
| 4.3 Mixed Integer Linear Programming | 51 |
| 4.4 Non-Linear Programming | 51 |
| 4.5 Quadratic Function | 52 |
| 4.6 Different Distributions | 53 |
| 4.6.1 Normal distribution | 53 |
| 4.6.2 Poisson distribution | 54 |
| 4.6.3 Gaussian distribution | 54 |
| 4.7 Dimension Reduction Methods | 55 |
| 4.7.1 Principal component regression (PCR) method | 55 |
| 4.7.2 Reduced rank regression (RRR) method | 55 |
| 4.8 Approximation Algorithms | 56 |
| 4.9 Summary | 57 |
| References | 57 |

# Part II
# Cloud Computing Applications for Smart Grid

## 5. Demand Response — 61

|   |   |
|---|---|
| 5.1 Fundamentals of Demand Response and Challenges | 61 |
| 5.2 Different Demand Response Mechanisms | 62 |
| 5.2.1 Economic demand response | 63 |
| 5.2.2 Emergency demand response | 65 |
| 5.2.3 Ancillary demand response | 66 |
| 5.3 Problems with Existing Approaches without Cloud | 66 |
| 5.4 Cloud-Based Demand Response in Smart Grid | 67 |

|  |  |
|---|---|
| 5.4.1 Demand response in smart grid energy management | 68 |
| 5.4.2 Demand response in data centers for the smart grid | 74 |
| 5.5 Future Trends and Issues | 77 |
| 5.6 Summary | 78 |
| References | 79 |

## 6. Geographical Load-Balancing — 81

|  |  |
|---|---|
| 6.1 Need for Load-Balancing in Smart Grid | 81 |
| 6.2 Challenges | 82 |
| 6.3 Problems with Existing Load-Balancing Approaches without Cloud | 83 |
| 6.3.1 Coalition formation | 83 |
| 6.3.2 Flexible demand forecasting | 83 |
| 6.3.3 Centralized load controller | 84 |
| 6.4 Cloud-Based Load-Balancing | 84 |
| 6.4.1 Price-based energy load-balancing | 85 |
| 6.4.2 Load-balancing at the smart grid data centers | 86 |
| 6.4.3 Renewable energy-aware load-balancing | 89 |
| 6.4.4 Load-balancing at data center networks | 90 |
| 6.5 Future Trends and Issues | 92 |
| 6.6 Summary | 93 |
| References | 94 |

## 7. Dynamic Pricing — 96

|  |  |
|---|---|
| 7.1 Deployment of Dynamic Pricing in Smart Grids | 96 |
| 7.1.1 Determination of actual time-slot | 96 |
| 7.1.2 Need for adequate infrastructure | 97 |
| 7.2 Existing Dynamic Pricing Policies without Cloud | 97 |
| 7.2.1 Day-ahead pricing policy | 97 |
| 7.2.2 Demand-based pricing policy | 98 |
| 7.2.3 Supply-based pricing policy | 99 |
| 7.2.4 Supply–demand-based pricing policy | 99 |
| 7.3 Problems with Existing Approaches without Cloud | 100 |
| 7.3.1 Local knowledge of supply–demand information | 100 |
| 7.3.2 Unfair pricing tariffs for customers | 100 |
| 7.4 Cloud-Based Dynamic Pricing Policies | 100 |
| 7.5 Future Trends and Issues | 103 |

|  |  |
|---|---|
| 7.6 Summary | 104 |
| References | 104 |

## 8. Virtual Power Plant — 106

|  |  |
|---|---|
| 8.1 Concept of Virtual Power Plant | 106 |
|     8.1.1 Commercial VPP | 108 |
|     8.1.2 Technical VPP | 108 |
| 8.2 Advantages of Virtual Power Plant | 109 |
|     8.2.1 Acts as internet of energy | 109 |
|     8.2.2 Energy efficiency | 110 |
|     8.2.3 Online optimization platform | 110 |
|     8.2.4 Systems security | 111 |
| 8.3 Virtual Power Plant Control Strategy | 111 |
| 8.4 Virtual Power Plant: Different Methodologies | 111 |
|     8.4.1 Integration of electric vehicles | 112 |
|     8.4.2 Implementation of energy storage devices | 113 |
| 8.5 Future Trends and Issues | 115 |
| 8.6 Summary | 116 |
| References | 116 |

## 9. Advanced Metering Infrastructure — 118

|  |  |
|---|---|
| 9.1 Requirements | 118 |
| 9.2 Different Approaches of AMI | 119 |
|     9.2.1 Data collection for AMI | 119 |
|     9.2.2 Classification of AMI data | 126 |
|     9.2.3 Security for AMI | 127 |
|     9.2.4 Electricity theft detection in AMI | 139 |
| 9.3 Future Trends and Issues | 142 |
| 9.4 Summary | 144 |
| References | 145 |

## 10. Cloud-Based Security and Privacy — 147

|  |  |
|---|---|
| 10.1 Security in Data Communication | 147 |
|     10.1.1 Overview of PKI | 150 |
| 10.2 Security and Privacy Challenges and Opportunities | 153 |
| 10.3 Security and Privacy Approaches without Cloud | 155 |
| 10.4 Cloud-Based Security and Privacy Approaches | 159 |

|  |  |
|---|---|
| 10.4.1 Identity-based encryption | 160 |
| 10.5 Future Trends and Issues | 163 |
| 10.6 Summary | 163 |
| References | 164 |

# Part III
# Smart Grid Data Management and Applications

## 11. Smart Meter Data Management — 169

- 11.1 Smart Metering Architecture — 169
- 11.2 Challenges and Opportunities — 170
  - 11.2.1 Requirement of scalable computing facility — 170
  - 11.2.2 Presence of heterogeneous data — 171
  - 11.2.3 Requirement of large storage devices — 171
  - 11.2.4 Information integration from different levels — 172
  - 11.2.5 Complex architecture in the presence of multiple parties — 172
- 11.3 Smart Meter Data Management — 172
  - 11.3.1 Cluster-based management — 172
  - 11.3.2 Data compression and pattern extraction — 176
  - 11.3.3 Cloud computing for big data management — 179
  - 11.3.4 Calibration with big data — 183
  - 11.3.5 Fuzzy logic-based management — 184
- 11.4 Future Trends and Issues — 186
- 11.5 Summary — 186
- References — 187

## 12. PHEVs: Internet of Vehicles — 189

- 12.1 Convergence of PHEVs and Internet of Vehicles — 190
- 12.2 Electric Vehicles Management — 191
  - 12.2.1 Charging and discharging of PHEVs — 191
  - 12.2.2 Energy management for data centers and PHEVs — 199
  - 12.2.3 Providing on-board internet service facility — 201
- 12.3 Future Trends and Issues — 204
- 12.4 Summary — 205
- References — 206

## 13. Smart Buildings — **207**

13.1 Concept of Smart Building — 207
13.2 Challenges and Opportunities — 208
13.3 Different Approaches for Establishing Smart Buildings — 209
    13.3.1 Automatic energy management systems — 209
    13.3.2 Intelligent information management systems — 217
13.4 Future Trends and Issues — 222
13.5 Summary — 223
    References — 224

# Part IV
# Smart Grid Design and Deployment

## 14. Simulation Tools — **227**

14.1 Simulation Tools — 227
    14.1.1 Open DSS — 227
    14.1.2 MATPOWER — 228
    14.1.3 NS-2 and NS-3 — 228
    14.1.4 GridSim — 229
    14.1.5 OMNeT++ — 229
    14.1.6 GridLAB-D — 229
    14.1.7 SUMO — 230
14.2 Summary — 230
    References — 230

## 15. Worldwide Initiatives — **231**

15.1 Initiatives Taken by EU — 232
15.2 Initiatives Taken by US Department of Energy — 237
15.3 Smart Grid Initiatives in Other Countries — 238
    15.3.1 Initiatives in China — 239
    15.3.2 Initiatives in India — 239
15.4 Smart Grid Standards — 241
15.5 Summary — 242
    References — 243

*Index* — 245

# Figures

| | | |
|---|---|---|
| 1.1 | Schematic view of smart grid | 4 |
| 1.2 | Smart grid framework by NIST | 6 |
| 1.3 | Smart grid communication model | 7 |
| 1.4 | Different peak periods in smart grid | 9 |
| 1.5 | Schematic view of different layers in the smart grid architecture | 11 |
| 1.6 | Smart grid interoperability designed by GWAC | 15 |
| 1.7 | Cross-cutting issues in smart grid | 16 |
| 2.1 | Overview of cloud computing technology | 19 |
| 2.2 | Abstracted view of sharing and procuring services through a cloud platform | 20 |
| 2.3 | Example of mainframe workloads: Batch job and online transaction | 22 |
| 2.4 | Schematic view of mainframe world | 22 |
| 2.5 | Schematic view of client–server architecture | 23 |
| 2.6 | Schematic view of a two-tier client–server architecture | 24 |
| 2.7 | Schematic view of a three-tier client–server architecture | 24 |
| 2.8 | Example of cluster computing with a load-balancer | 25 |
| 2.9 | Schematic view of grid computing architecture | 27 |
| 2.10 | Schematic view of service oriented architecture | 27 |
| 2.11 | Example of hardware virtualization | 29 |

| | | |
|---|---|---|
| 2.12 | Schematic view of KVM architecture | 30 |
| 2.13 | Different services, models, and properties of cloud computing | 32 |
| 2.14 | Schematic view of the cloud deployment models | 33 |
| 2.15 | Cloud computing architecture | 36 |
| 3.1 | Growth of information storage capacity with digitization | 39 |
| 3.2 | Different attributes of big data | 40 |
| 3.3 | Life cycle of data defined by CRISP | 42 |
| 3.4 | Benefits of big data analytics | 45 |
| 3.5 | Benefits of big data analytics in the smart grid | 46 |
| 4.1 | Simplified view of a normal distribution | 53 |
| 5.1 | Cloud-based demand response model for the smart grid | 68 |
| 5.2 | Two-layer hierarchical cloud-based demand response | 69 |
| 5.3 | Micro-grid providing electricity to customers with different energy generators | 71 |
| 5.4 | Two-stage cloud-based demand response model | 72 |
| 5.5 | Electric demand response in data centers | 76 |
| 6.1 | Service pooling and assigning for load-balancing in data centers | 86 |
| 6.2 | Schematic view of a data center powered by smart grid | 87 |
| 6.3 | Schematic diagram of a transmission delay-based load-balancing using front-end servers | 88 |
| 6.4 | Schematic view of geographical load-balancing at data center network | 91 |
| 7.1 | Schematic view of cloud instance scheduler | 103 |
| 8.1 | Schematic diagram of a virtual power plant | 107 |
| 8.2 | Large-scale view of virtual power plant | 107 |
| 8.3 | Conceptual view of a commercial virtual power plant (CVPP) | 108 |
| 8.4 | Conceptual view of a technical virtual power plant (TVPP) | 109 |
| 8.5 | Layered architecture of different control strategies for the VPP | 111 |
| 8.6 | Agent-based energy storage devices management in VPP | 114 |
| 8.7 | Virtual integration of different energy management units | 114 |
| 8.8 | Dynamic energy dispatch through virtual integration | 115 |
| 9.1 | Steps for information collection | 120 |
| 9.2 | Header format used for information collection | 120 |

| | | |
|---|---|---|
| 9.3 | Rank-based routing scheme for AMI architecture | 123 |
| 9.4 | Heuristic-based gateway selection: (a) A sample network topology is created with six smart meters; (b) Possible minimum spanning trees (equal cost for one-hop distance); (c) Obtained minimum distance matrix $M$, and the maximum hop-distance $L$. According to the scheme, C is selected as the gateway node to optimize the hop-distance to all SMs | 125 |
| 9.5 | Steps for securing an AMI system | 128 |
| 9.6 | Schematic diagram of a smart meter enabled with IDS | 129 |
| 9.7 | Schematic view of key management framework by forming a key graph | 130 |
| 9.8 | Group-based smart grid communication | 135 |
| 9.9 | Merkle tree for eight revoked certificates | 137 |
| 9.10 | Certificate verification process | 138 |
| 9.11 | Verification of a certificate to check whether or not it is revoked | 139 |
| 9.12 | Flowchart for theft detection model | 143 |
| 10.1 | Different security aspects of smart grid | 154 |
| 10.2 | Schematic view of cloud-based secure data communication in smart grid | 160 |
| 11.1 | Schematic view of the smart meter information collection with different layers | 170 |
| 11.2 | Energy consumption profile for two different customers in smart grid | 174 |
| 11.3 | Discrete representation of continuous time series data | 175 |
| 11.4 | Schematic view of the identity-based cryptosystem | 181 |
| 11.5 | Schematic view of an identity-based signature | 182 |
| 11.6 | Hierarchical levels of identity-based encryption with signature | 182 |
| 11.7 | Big data for distribution parameter estimation framework | 183 |
| 11.8 | Algorithm for parameter estimation | 184 |
| 11.9 | Framework for outage management in the smart grid (adopted from [26]) | 185 |
| 12.1 | Different services that can be obtained from a PHEV | 191 |
| 12.2 | Schematic view of aggregator-controlled PHEVs in a smart grid | 192 |
| 12.3 | Coordinated charging/discharging of PHEVs while forming different groups | 195 |
| 12.4 | Different actions to charge/discharge the PHEVs | 196 |
| 12.5 | Schematic view of resource reservation in a smart grid in the presence of PHEVs | 198 |
| 12.6 | Schematic system model of energy regulation with data centers and PHEVs | 199 |

12.7 Difference between VANET and information-centric networking 203
12.8 Integrated architecture of the internet of vehicles and clouds 204
13.1 Schematic view of WSN-based smart building system 210
13.2 Working flow of energy source scheduling in a smart building 211
13.3 Flow chart for rule verification process in a smart building 213
13.4 Architecture for appliance control inside a smart building 214
13.5 Appliances finite state machine 215
13.6 Cluster formation inside a smart building 218
13.7 Packetized view of energy supply in the smart grid 220
13.8 Cloud-based smart home management 222
15.1 Schematic overview of smart grid network according to ETP 233
15.2 H1 and H2 interfaces used in smart grid communication 236
15.3 Steps to address interoperability issues in smart grid 236

# Tables

| | | |
|---|---|---|
| 1.1 | Basic differences between power grid and smart grid | 5 |
| 5.1 | Comparison of different demand response approaches | 77 |
| 6.1 | Comparison of different load-balancing approaches | 92 |
| 10.1 | Security features in data communication in smart grid | 161 |
| 15.1 | Smart grid deployment status in Europe | 232 |
| 15.2 | Distribution of projects in European countries | 234 |
| 15.3 | Smart grid deployment status in the US | 237 |
| 15.4 | Smart grid investment on grant projects | 238 |
| 15.5 | Smart grid initiatives in India | 240 |
| 15.6 | Smart grid standards and their brief description | 241 |

# Foreword

The growing concern about climate change and energy conservation necessitates the existing power grid to be modernized. Consequently, bi-directional communication network is placed atop the existing power grid, named as smart grid, with an aim to reduce energy loss and to provide improved energy management. This makes smart grid a cyber-physical system with a strong intertwining between the software, communication, control, and physical components. The cyber-physical property of smart grid necessitates adequate energy and information management, while considering the associated security concerns. Further, millions of smart meters are expected to be deployed to collect real-time status of energy consumption, and the collected information should be processed in real-time. Consequently, smart grid requires support of advanced computing platform and data management tools to manage real-time energy requirements and to process real-time information, respectively. Cloud computing and big data technologies are the topics of interest among the scientific community due to their inherent features of advanced computation and data management.

Unlike most smart grid books, Sudip Misra and Samaresh Bera in this book focus on the specific challenges that are presented in the smart grid while integrating multiple entities together, which can be potentially solved using cloud computing technology and advanced data management approaches. It covers main aspects of smart grid, touching energy management, information management, and security, by leveraging the advantages of cloud computing technology and advanced data management approaches. The book discusses the important issues, such as demand response, dynamic pricing, load balancing, and smart meter data management, and includes the treatment to deal with such issues. In brief, the attractive features of this book are catalogued below:

a) Fundamental concepts of smart grid, cloud computing, and big data analytics are discussed. Additionally, the commonly used mathematical models and tools in this domain are also given concisely.
b) Cloud computing technology for smart grid presents a comprehensive discussion on the benefits of cloud computing technology in the smart grid for improved energy management and security.
c) Data management approaches mainly focus on the information management in the smart grid.
d) Smart grid simulation tools and initiatives worldwide highlight the recent efforts and on-going projects to realize the smart grid technology in a large-scale.

While the feature (a) would aid in the understanding of the basic concepts for novice readers such as students in a first course on smart grid, (b), (c), and (d) are expected to serve as assets for both researchers and practitioners alike.

The authors of the book, Sudip Misra and Samaresh Bera, are globally acclaimed researchers, both of whom have published a number of research papers in this domain in respectable journals such as the IEEE Transactions on Smart Grid, IEEE Transactions on Parallel and Distributed Systems, Elsevier Computer Networks, and IEEE Systems Journal. Sudip is well-known in the community for his research achievements in the broad domain of Internet of Things. He has published close to a dozen books, which are published by the Cambridge University Press, Wiley, Springer, World Scientific, and CRC Press. Due to his significant research contributions, his works were recognized with different fellowships and awards such as the Fellow of the National Academy of Sciences (India), IEEE Communications Society Outstanding Young Researcher Award, Humboldt Fellowship (Germany), Faculty Excellence Award (IIT Kharagpur), Canadian Governor Generals Academic Medal, NASI Young Scientist Award, IEI Young Engineers Award, SSI Young Systems Scientist Award and so on. He serves as the Associate Editor of the IEEE Transactions on Mobile Computing and the IEEE Transactions on Vehicular Technology. The second author Samaresh is a senior researcher in the Smart Wireless Applications and Networking (SWAN) Lab at IIT Kharagpur. He was awarded the IEEE Richard E. Merwin scholarship by IEEE Computer Society for his outstanding contributions to IEEE. He also won the Young Researcher Award in the Heidelberg Laureate Forum (Germany).

In closing, I would like to stress that the book is expected to be a rich resource for students, researchers, and practitioners, written by well-known researchers in the community. The material available in the book is a mixture of theory and applications, and is structured in an easily readable form.

I trust the readers will enjoy reading this book!

Professor Abbas Jamalipour, PhD
Fellow of IEEE, Fellow of IEICE, Fellow Engineers Australia
Head of School and Professor of Ubiquitous Mobile Networking
The University of Sydney, Australia

January 2018

# Preface

## Overview and Goals

Due to the growing concerns about climate change and energy conservation, there has been a greater focus than ever before on the development of digitalized and green technology for modern power delivery systems – Smart Grid – to the customers. Smart grid is conceptualized as the integration of bi-directional communication networks on the existing electricity networks. With the help of this type of communication networks, service providers and customers have real-time information about the energy consumption and price of energy, respectively. Therefore, service providers can maintain a balance between real-time supply and demand, so as to maximize their revenues. On the other hand, customers can consume energy in an optimized manner to minimize their energy consumption cost. Consequently, different intelligent mechanisms – demand response, dynamic pricing, integration of renewable energy sources, and distributed energy service – are introduced to support the smart grid architecture.

Concurrently, cloud computing is an emerging technology that facilitates on-demand, real-time service platform dynamically with shared services to its clients. Cloud computing primarily provides three different services: software as a service (SaaS), infrastructure as a service (IaaS), and platform as a service (PaaS). Therefore, clients can utilize cloud resources according to their individual requirements. Another important feature of cloud computing is that it supports the implementation of a distributed architecture.

Recently, different studies have revealed that cloud computing is a promising technology that can support the smart grid requirements from different perspectives: energy management, information management, and security services. Real-time energy management is the most important feature of the smart grid. Different schemes are introduced for this purpose: demand scheduling, integration of storage devices, cooperative architecture, are a few examples. A micro-grid in smart grid provides electricity to customers in a distributed manner with the help of renewable and non-renewable energy sources. Therefore, all entities (such as customers, service providers, and third parties) require adequate real-time information in order to execute the said schemes. Consequently, it is required to have a common platform from where all the entities can fetch adequate real-time information, which, in turn, necessitates the availability of suitable information management schemes. In the smart grid, a massive number of smart meters is expected to be deployed in order to have real-time information about customers' energy consumption patterns. Presently, big data management approaches have been popular for processing and storage of massive volumes of heterogeneous data generated at high speeds. Therefore, intelligent data management is also required for smart grid, in order to take appropriate decisions. On the other hand, allowing access to the customers' energy consumption pattern may lead to a security breach in the smart grid. Therefore, security and privacy of customers should be preserved along with information management. In such a scenario, cloud computing and big data technologies are suitable supplements to smart grid, in order to support real-time energy management in a cost-effective, reliable, and secure manner.

This book discusses different well-known smart grid technologies, their concepts and underlying principles, from the perspective of integration with cloud computing and data management approaches. It will be helpful to students, researchers, and practitioners to get a thorough insight into the existing cloud and data analytics-based smart grid technologies and their shortcomings in practical environments. Different cloud and big data-based technologies are discussed from three different perspectives: energy management, information management, and security in the smart grid. Therefore, the book will be helpful to a wide range of audience in their respective domains in the smart grid.

Specifically, this book will be useful to graduate students, researchers, and practitioners working in different fields spanning of computer and electrical sciences both as a reference book and as a text book.

## Organization and Features

The book is broadly divided into three parts. Part I consists of three chapters. Chapter 1 presents the preliminary concepts of smart grid. Chapter 2 is dedicated to the basic concepts of cloud computing. Chapter 3 discusses the different concepts of big data analytics. Chapter 4 highlights the fundamental mathematical techniques/concepts that are useful in modeling and analyzing the smart grid system. Moreover, the mathematical tools that are prerequisites to understanding the problems and their solutions in the subsequent chapters are listed.

Part II, which comprises six chapters, focuses on the recent advances in cloud computing applications in smart grid. Chapter 5 discusses different demand response mechanisms in the smart grid using cloud computing approaches. Chapters 6 and 7 discuss different load balancing approaches using geographical locations and dynamic pricing in smart grid, respectively, from the perspective of cloud computing. Chapters 8 and 9 focus on integration of virtual energy systems and advanced metering infrastructure to collect smart meter information, respectively. Finally, Chapter 10 discusses different security and privacy policies in the smart grid using cloud computing approaches.

Part III consists of three chapters, and is dedicated to the advanced topics in smart grid data management, in which different problems can be potentially solved using different data management approaches. Chapter 11 discusses the suitability of PHEVs as connecting devices of the different entities of the smart grid. Chapters 12 and 13 focus on smart meter data management and smart building management in the smart grid, respectively.

Part IV consists of two chapters, and focuses on the designing and experimenting tools of smart grid architecture. Further, it also focuses on the current deployment status of smart grid technology worldwide. A list of smart grid simulation tools is presented to provide an overview of the designing and modeling smart grid architecture in Chapter 14. Chapter 15 highlights the smart grid initiatives taken worldwide.

We list some of the important features of this book, which, we believe, make this book a valuable resource to our readers.

- Description of research work integrating cloud computing and big data applications for smart grid.
- Description of how and why cloud computing and big data analytics are useful for smart grid technology.
- A balanced mixture of theory and applications.
- Highlighting of smart grid simulation tools
- Highlighting of the current status of smart grid development
- Extensive bibliography
- Sample questions

## Target Audience

This book is written for the student community at all levels: from those new to the field to undergraduates and postgraduates. It can be used for advanced elective courses on smart grid, cloud, and big data, as well.

The secondary audiences for this book are the research communities in both academia and industry. To meet the specific needs to these groups, most chapters include sections discussing directions of future research.

Finally, we have considered the needs of readers from the industries who seek practical insights: for example, how cloud computing applications and big data analytics are useful in real-life smart grid systems.

# Part I
# Introduction

# CHAPTER 1

# Introduction to Smart Grid

A smart grid is conceptualized as the existing power grid supported by bi-directional communication networks. Therefore, with the help of communication networks, service providers have real-time information about energy supply and demand. Moreover, customers also have real-time price information, based on which they can consume energy in an *optimized* manner, so that the total energy consumption cost is reduced. In addition to communication networks, several distributed energy generation units (such as solar, wind, and combined heat power) are also integrated to deploy distributed energy management policies. These distributed units are known as *micro-grids*. Typically, a smart grid consists of the following components – micro-grid, smart meter, renewable energy sources, and plug-in hybrid electric vehicles (PHEVs) [1]. Figure 1.1 depicts a schematic view of the smart grid architecture. Table 1.1 presents the basic differences between the traditional power grid and the smart grid.

## 1.1 Smart Grid Framework and Communication Model

Smart grid framework was initially conceptualized by the National Institute of Standards and Technology (NIST) in 2009. Figure 1.2 shows the smart grid framework conceptualized by NIST in its framework version 1.0 (adopted from [2]). The smart grid framework includes several entities – operators, markets, customers, utility providers, generation units, transmission units, and distribution units. The role of operators includes monitoring of the overall smart grid system and implementation of policies, while considering the interests of other entities. On the other hand, utility providers act as middle-persons between the customers and the energy markets. Therefore, it is possible to

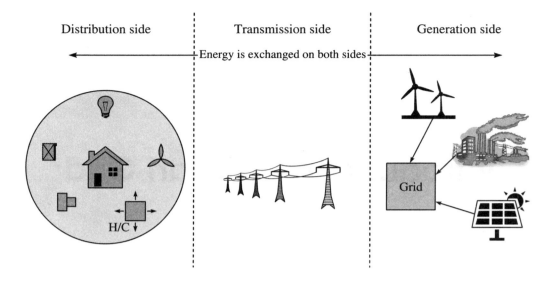

**Distribution side** includes local substations and customers. Further, the local substations are built with modern technologies to integrate energy from the main power grid with energy generated locally at the substation in a distributed manner. Therefore, the power supply to the customers is done in a semi-distributed manner. The local substation reports to the main power grid about the local energy generation and demand from the customers.

Customers' homes are equipped with 'smart meters,' which send the energy consumption report to the substations. Further, the appliances installed at the customers' ends are connected to the smart meters, in order to have smart energy monitoring system.

**Transmission side** mainly focuses on the power delivery to the distribution side from the generation side. The transmission side is also enabled with bi-directional communication facility, in order to have real-time status of the generation and transmission sides. Typically, power line communication (PLC) is used to exchange such information. However, presently, multiple communication technologies are placed for improved smart grid monitoring.

**Generation side** focuses on the power generation. which includes generation from traditional power plants (for example, the power plant based on fossil fuels) and hybrid systems (for example, combined heat power). Additionally, wind power and solar power are also integrated at the generation side. However, the intermittent behavior of wind and solar power must be considered while generating energy from such sources. The generation side is also equipped with modern communication technologies to have real-time energy supply status.

**Figure 1.1** Schematic view of smart grid

have multiple utility providers in a single energy market. The energy generation, transmission, and distribution units are responsible for generating, transmitting, and distributing energy supply to the customers, respectively, while considering the status of the energy market. Further, NIST revised the framework while considering other entities that are important in a smart grid energy management system.

Table 1.1   Basic differences between power grid and smart grid

| Characteristics | Power Grid | Smart Grid |
|---|---|---|
| Customer participation | Static policies are deployed irrespective of real-time energy consumption from customers. | Dynamic policies are expected to be deployed while considering the real-time energy consumption from customers. |
| Communication facility | Only power line communication (PLC) is present. | Bi-directional communication facility is present, in which licensed and unlicensed frequency bands are in use. |
| Distributed energy generation | Few distributed generations such as solar and wind energy are considered. | All types of distributed generations such as solar and wind energy, and combined heat power (CHP) are taken into consideration. |
| Inclusion of storage devices | Not present. | Different distributed energy storage devices, which can be used in different situations are considered. |
| Real-time consumption monitoring | Not present. | Using bi-directional communication facility, real-time energy consumption can be monitored. |
| Security | No proper implementation is present to prevent energy theft and other types of security breach. | Adequate policies are taken into consideration to prevent energy theft and other types of security breach. |

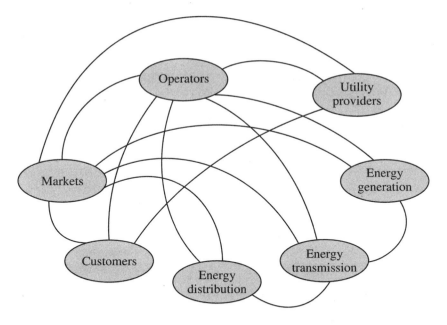

**Figure 1.2** Smart grid framework by NIST

Concurrent to the overall framework, the smart grid communication model is also conceptualized for information collection and management in the smart grid. Figure 1.3 presents the smart grid communication model comprising of backbone networks, data center networks, access points, and end-devices. The backbone network is responsible for routing real-time information among all smart grid entities. On the other hand, data center network stores the information, and processes it to take adequate decisions in order to have reliable and cost-effective energy management in the smart grid. The access point is responsible for information collection from the end-devices. Additionally, it also acts as an aggregator in the smart grid system. The end-devices correspond to smart meters installed at the customers' end. Further, the smart meters may be connected to the appliances installed in homes.

## 1.2 Smart Grid Vision

According to the Department of Energy [6], a smart grid is envisioned to fulfill the following objectives:

- Operational efficiency: One of the objectives is operational efficiency. Further, this corresponds to the following sub-objectives:
  - Integration of distributed generation units
  - Improvement of resource utilization

Introduction to Smart Grid 7

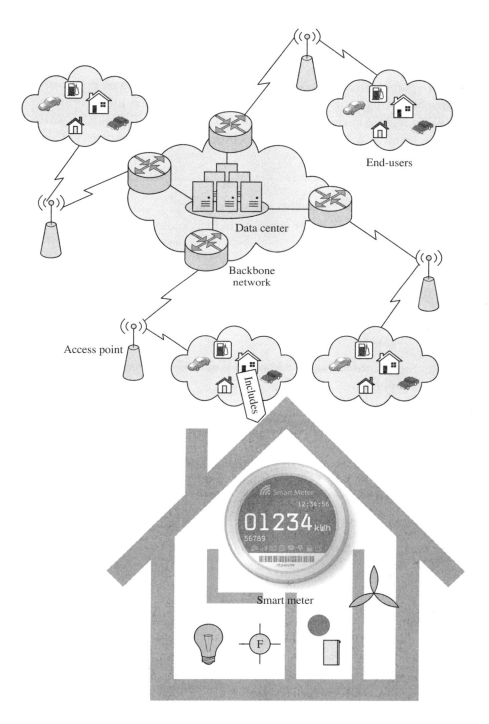

**Figure 1.3** Smart grid communication model

- Real-time energy supply–demand monitoring
- Intelligent system

- Active participation of customers: In the traditional power grid, customers do not actively participate in the energy trading process. The utility providers define the policies, and the customers are required to follow these. In contrast, active customers' participation is one of the primary objectives of the smart grid. This includes the following:
  - Reduction in electricity outage
  - Engagement of customers in energy trading
  - Policy enforcement for active participation of customers

- Energy efficiency: Energy efficiency is mainly focused on reliable energy distribution to the customers. This includes the following:
  - Reduction in energy loss
  - Reduction of imbalance between supply and demand
  - Reliable energy service

- Green energy: Reduction in carbon emissions in the energy generation, transmission, and distribution.

## 1.3 Requirements of a Smart Grid

The growing development of smart grid systems necessitates different requirements to be fulfilled. Sections 1.3.1–1.3.5 identify some of the most important ones.

### 1.3.1 Energy management

Real-time energy management is the main objective of the smart grid. The existing power grids require optimal balance between the real-time energy supply from all generation units and demand from all customers. Various schemes such as home energy management (HEM) [5, 6], building energy management (BEM) [4], and demand side management (DSM) [7] are introduced in the smart grid system to fulfill these requirements. In home energy management, specifically, an energy management unit (EMU) is installed inside the home; this unit monitors real-time energy consumption by the appliances available within the home. Based on the real-time price signal, the EMU optimally manages the appliances, so that energy consumption during peak hours can be avoided. Typically, the entire time period is categorized as *off-peak*, *mid-peak*, and *on-peak* periods. During off-peak periods, total energy supply to the grid is higher that the total energy demand from customers. Energy supply and demand are moderate during mid-peak periods. In contrast, during on-peak periods, energy demand from customers is higher than the energy supply to the grid. Figure 1.4 depicts three different time periods in a smart grid environment. Depending on the real-time energy supply–demand situation, intelligent units take decisions to balance the energy supply and demand. Adequate balance between the energy supply and demand is required in order to have an optimized energy management in the smart grid.

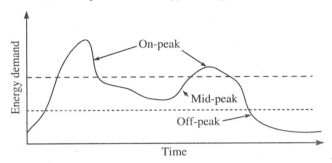

**Figure 1.4** Different peak periods in smart grid

### 1.3.2 Need to support multiple devices

In a smart grid, three major components exist – the generation side, the transmission side, and the distribution side, as depicted in Figure 1.1. In the generation side, multiple energy generators (such as renewable and non-renewable) are expected to be present, which are further be equipped with multiple devices such as sensors and actuators. For example, the renewable energy sources are solar and wind energy. Sometimes they are also known as non-dispatchable energy sources, as they cannot be generated according to the requirements. On the other hand, an example of a non-renewable energy source is fossil fuel. Sometimes, these are also known as dispatchable energy sources, as energy from such sources can be generated according to the requirements. As they can be generated according to requirements, non-renewable energy sources are more reliable than the renewable ones. However, the former has higher carbon footprint. Energy is supplied to the distribution side from the transmission side. The supplied energy is distributed through the distribution (power) networks. Consequently, multiple devices are placed at each of the components to monitor the real-time energy supply–demand status, in order to have a balanced smart grid environment. Hence, it can be noted that multiple heterogeneous devices operate in a common platform, for which it is required to have adequate infrastructure facility.

### 1.3.3 Information management

The modern-day power grid is supported by bi-directional communication networks, referred to as the smart grid. Multiple communicating devices also participate in addition to the traditional appliances. Consequently, the data generated from the implanted sensors and actuators needs to be managed in an adequate manner. Existing information management schemes are required to be revised considering the requirements of the smart grid environment. For example, appliances may communicate with the home gateways using Zig-Bee (based on IEEE 802.15.4 protocol) or Wi-Fi (based on IEEE 802.11 protocol) networks. Therefore, the packet formats for different communication protocols are different, which need to be taken into account while aggregating the information at the gateways. Furthermore, multiple consumers (such as homes, office buildings, and shopping malls) have different energy consumption profiles. Therefore, adequate pricing and billing policies are also required to be implemented for different types of consumers, which, in turn, requires adequate information management for different consumers.

### 1.3.4 Layered architecture

The framework of a smart grid consists of multiple layers from two different perspectives – energy and communication. The layered architecture of the energy consists of multiple energy layers through which energy is managed in the smart grid. On the other hand, using the communication layered architecture, bi-directional information is managed. Figure 1.5 presents a schematic view of different layers existing in a smart grid. As depicted in Figure 1.5, the energy and communication layers consist of the distribution, transmission, and generation layer. In the energy layers, different energy consumption, transmission, and generation units are presented. On the other hand, in the communication layer, different communication technologies, which have the potential to support smart grid requirements, are present. For short-range communication, technologies such as Zig-Bee, Bluetooth, and Wi-Fi, are used. In contrast, for long-range communication, WiMax or GSM may be used at the transmission layer. Consequently, such layered architecture must be maintained and supported in order to establish a smart grid environment for improved energy supply demand management.

### 1.3.5 Security

Security is crucial to the success of smart grid. As evident from the aforementioned points, multiple devices and parties participate in smart grid energy management. Therefore, adequate security mechanisms must be employed to secure energy information of different entities. For example, one customer should not be able to access the energy consumption profile of another, without following the proper authentication procedure. Similarly, access to different information can be maintained in a layered architecture. For example, a substation can monitor its own customers. However, it cannot monitor the entire energy consumption profile of other substations within the smart grid environment. In contrast, the utility provider can access information of all substations operating under it. The existing security mechanisms can be applied. However, due to the resource constrained nature of the smart grid entities, existing security schemes may not be adequate. Moreover, over millions of customers may participate in the energy trading process in real-time. Therefore, issue of key sharing is an important concern. Consequently, these issues/requirements should be addressed prior to large-scale deployment of the smart grid environment.

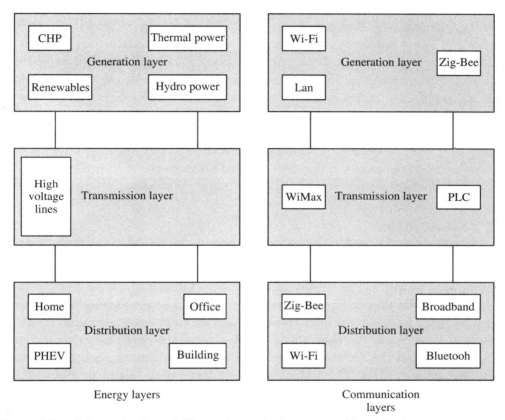

**Figure 1.5** Schematic view of different layers in the smart grid architecture

## 1.4 Components of the Smart Grid

To support different requirements of the smart grid, several components were introduced by researchers during the past years. Examples include smart meter, bi-directional communication network, micro-grid, plug-in hybrid electric vehicles, and renewable energy sources.

### 1.4.1 Bi-directional communication

Bi-directional communication network is the main backbone of a smart grid. Integration of bi-directional communication converts the traditional power grid into the modern-day smart grid. Through the bi-directional communication network, the real-time energy status is monitored at both the customers' and utility providers' ends. Depending on the real-time information availability to the entities, adequate decisions are taken so that an efficient energy management system can be deployed. Different communication protocols are expected to be used, such as IEEE 802.15.4, IEEE 802.11, and IEEE 802.16, as mentioned before. As presented in Figure 1.5, IEEE 802.15.4-based communication technology, such as Zig-Bee, can be used to establish communication between home gateway and home appliances. Consequently, the home appliances can be controlled through the gateway devices, and energy consumption at the appliances can be reported to the gateway devices using the Zig-Bee communication technology. On the other hand, the home gateways can communicate with the local distribution network using the IEEE-802.11-based communication technologies. Finally, long-range communication can be established using the IEEE 802.16-based communication technologies. Therefore, with the help of a bi-directional communication network, it is possible to keep track of the real-time status of the energy supply and demand information. Accordingly, the service providers can manage the power network in a more efficient manner.

### 1.4.2 Smart meter

Smart meter is another important component of the smart grid. Typically, smart meters are installed at customers' premises to record energy consumption information. Moreover, they support bi-directional communication, through which the utility providers and customers exchange information. Thus, real-time energy monitoring and billing policies are made easier for the utility provider with the help of the smart grid. Apart from monitoring home energy consumption, smart meters also help in gathering energy consumption data at the distribution level, and thus, report remote energy consumption data to the utility providers. Building advanced metering infrastructure (AMI) is an important application of smart meters in smart grid environment. The concept of AMI will be discussed in Chapter 9.

Communication technology is the main additional feature of a smart meter. Unlike traditional power line communication (PLC) in an electric distribution system, a smart meter uses dedicated protocols to communicate with the utility providers. Different communication protocols are proposed and are being used in the smart meters in order to have real-time energy management. As discussed earlier, Wi-Fi is a promising technology, which can be used with smart meters to communicate with the local distribution networks. IEC 62056 [8] is the most widely used communication standard in smart meters for data exchange. Smart meter data are sent using serial ports in the form of ASCII code. However, adequate security mechanism is required to protect smart meter data. Currently, IEC 62056 uses different encryption methods, such as message authentication code (MAC) or signature-based algorithm (RSA), in order to secure smart meter data. Additionally, embedded intrusion detection systems can also be included within a smart meter, which is discussed in detail in Chapter 9.

## 1.4.3 Micro-grid

A micro-grid is conceptualized as a group of local distributed electricity generators that adds to the traditional centralized electric grids. Therefore, a micro-grid consists of several distributed generation and renewable energy sources (such as solar and wind). Consequently, it distributes electricity to the customers as a combination of distributed energy and centralized energy supplied from the main grid. As a result, a micro-grid maintains the balance between energy supply and demand when there is any surplus/deficit of real-time energy in the main grid. More importantly, a micro-grid can operate in an islanded mode when there is a problem in the main grid. The micro-grids can also be treated as local substations with some distributed generation facility. When there is adequate distributed energy supply to satisfy real-time energy demand from the customers. A micro-grid can also act in an islanded mode. On the other hand, a micro-grid buys electricity from the main grid when there is a deficit in energy supply generated from the distributed energy generators. Further, a micro-grid sells energy back to the main grid when there is a surplus in energy supply from the distributed generators. The main components of a micro-grid are as follows:

- Local generation: A micro-grid consists of several local generated units to meet customers' energy demand in real-time. Some of the main generated units are as follows: dispatchable energy sources (such as fossil fuel and combined heat power) and non-dispatchable energy sources (such as solar and wind power). The dispatchable and non-dispatchable energy sources are also known as non-renewable and renewable energy sources, respectively. In the dispatchable energy sources, traditional energy sources are used from which energy can be supplied depending on the real-time requirements. In contrast, energy cannot be generated depending on the requirements from non-dispatchable energy sources, as such energy sources depend on the environmental parameters. Additionally, the supplied energy from non-dispatchable energy sources are intermittent in nature, which needs to be considered. However, non-dispatchable energy sources attract interest among the community due to their inherent features, such as the capacity of providing green energy. Several demand–response mechanisms, which are useful for managing real-time energy supply and demand in the presence of both dispatchable and non-dispatchable energy sources, are discussed in Chapter 5.

- Energy consumers: Besides multiple energy generators, a micro-grid also consists of multiple consumers, which use energy from it. Some of the major consumers are residential customers, buildings and offices, industries, and plug-in hybrid electric vehicles (PHEVs). It is noteworthy that, presently, consumers also have distributed renewable energy generators (mainly solar energy). Therefore, the consumers can also sell back energy to the micro-grid.

- Energy storage: As there are multiple renewable energy generators that are intermittent in nature, the generated energy from such sources needs to be stored.

As a result, we also have different energy storage units in a micro-grid. The generated energy from the renewable energy sources is stored in the storage devices. The stored energy is supplied when there is a deficit in real-time energy supply from other energy generators to meet real-time energy demand from customers.

- Coupling point: Finally, a coupling point controls the energy exchange between the micro-grid and the main grid. Depending on the real-time situation, energy can be supplied to the micro-grid from the main grid and vice-versa.

### 1.4.4 Plug-in hybrid electric vehicles

A plug-in hybrid electric vehicle (PHEV) is another important component of the micro-grid [9]. PHEVs consume energy from the micro-grid and can also sell back energy to the micro-grid. These are achieved through grid to vehicle (G2V) and vehicle to grid (V2G) technologies. In a G2V process, energy is consumed by the PHEVs to charge their batteries. As we have already seen, there are different periods of loads in the smart grid – off-peak, mid-peak, and on-peak. Consequently, PHEVs can be charged during off-peak hours, so as to relieve the energy demand from the micro-grid during on-peak hours. In contrast, using the V2G technology, PHEVs can sell back energy to the micro-grid during on-peak hours to meet the huge energy demand from residential consumers including offices and industries. Therefore, suitable energy management policies are employed to maintain the real-time energy supply–demand balance. Consequently, PHEVs play an important role in smart grid energy management in real-time. We will discuss the potential of PHEVs in the smart grid in detail in Chapter 12.

## 1.5 Smart Grid Interoperability

In terms of designing, the smart grid is categorized into multiple levels, as presented in Figure 1.6 (adopted from [3]). These levels should be able to interact with each other, which, in turn, poses the challenge of interoperability among the levels.

- Basic connectivity: This level focuses on the establishment of connectivity among devices within a network. For example, appliances installed inside a smart home are connected to a smart meter. Therefore, adequate communication protocols are required to establish connection among multiple devices within a network.

- Network interoperability: This level focuses on the establishment of connectivity among multiple devices from multiple networks. Therefore, interoperability between protocols is considered in this level. This also enables communication among multiple systems.

- System interoperability: Data structure should be similar in information exchanged among systems, so that required data can be extracted from received messages.

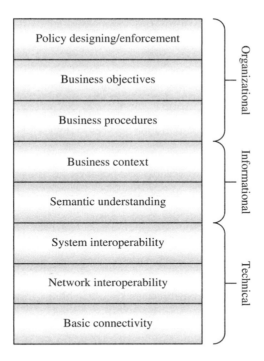

**Figure 1.6** Smart grid interoperability designed by GWAC

- Semantic interoperability: This level focuses on understanding the concepts of data structures of messages exchanged between systems.
- Business context: This level focuses on the awareness of smart grid technology from a business perspective.
- Business procedures: The relation between business processes and procedures are captured in this level.
- Business objectives: Business strategic objectives are decided in this level.
- Policy designing/enforcement: It focuses on the pricing policy and customers' engagement in the smart grid system.

According to a government-wide acquisition contract (GWAC) report [3], there exists few cross-cutting issues with the interoperability categories. Figure 1.7 presents the cross-cutting issues existing in smart grid.

## 1.6 Summary

In this chapter, the concept of smart grid was introduced. The requirements to establish a smart grid environment were presented. Finally, different components of a smart grid were discussed. In the subsequent chapters, several schemes will be discussed in detail to get an idea about smart grid energy and information management, while focusing on the security aspects of it.

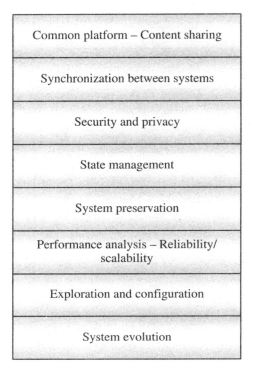

**Figure 1.7** Cross-cutting issues in smart grid

### Test Your Understanding

Q01. What is a smart grid?
Q02. What are the requirements of a smart grid?
Q03. State the various components of a smart grid.
Q04. What is a micro-grid?
Q05. What are the main objectives of a smart grid?
Q06. State the different time-periods widely used in smart grid.
Q07. What is a smart meter?
Q08. State the function of plug-in hybrid electric vehicles (PHEVs).
Q09. What is meant by dispatchable and non-dispatchable energy sources?

# References

[1] Bera, S., S. Misra, and Joel. J. P. C. Rodrigues. 2015. 'Cloud Computing Applications for Smart Grid: A Survey'. *IEEE Transactions on Parallel and Distributed Systems* 26 (5): 1477–1494.

[2] NIST Smart Grid Framework, Version 1.0. Accessed 05 September 2017. Available at ttps://www.nist.gov/sites/default/files/documents/public_affairs/releases/smartgrid_interoperability_final.pdf

[3] GridWise Architecture Council. 'GridWise Interoperability Context-Setting Framework'. Last modified 2008. Available at www.gridwiseac.org/pdfs/.

[4] Rocha, P., A. Siddiqui, and M. Stadler. 2015. 'Improving Energy Efficiency via Smart Building Energy Management Systems: A Comparison with Policy Measures'. *Energy and Buildings (Elsevier)* 88: 203–213.

[5] Misra, S., A. Mondal, S. Banik, M. Khatua, S. Bera, and M. S. Obaidat. 2013. 'Residential Energy Management in Smart Gird: A Markov Decision Process Based Approach'. In *Proc. of IEEE iThings/CPSCom*, pp. 1152–1157.

[6] Bera, S., P. Gupta, and S. Misra. 2015. 'D2S: Dynamic Demand Scheduling in Smart Grid Using Optimal Portfolio Selection Strategy'. *IEEE Transactions on Smart Grid* 6 (3): 1434–1442.

[7] Gelazanskas, L. and K. A. A. Gamage. 2014. 'Demand Side Management in Smart Grid: A Review and Proposals for Future Direction'. *Sustainable Cities and Society* 11: 22–30.

[8] IEC 62056. 2017. *Electricity Metering Data Exchange The DLMS/COSEM Suite*. Geneva, Switzerland: International Electrotechnical Commission.

[9] Misra, S., S. Bera, and T. Ojha. 2015. 'D2P: Distributed Dynamic Pricing Policy in Smart Grid for PHEVs Management'. *IEEE Transactions on Parallel and Distributed Systems* 26 (3): 702–712.

# CHAPTER 2

# Introduction to Cloud Computing

The term 'cloud computing' was initially coined to refer to on-demand computing services that were offered by commercial service providers such as Amazon, Google, and Microsoft. Cloud computing is an emerging computation model that shares resources over the Internet and provides on-demand facilities to its customers. In other words, it can be seen as a computation platform combined with shared computer processing resources over the Internet. Computing resources can be shared from anywhere using Internet connectivity. Thus, cloud computing enables ubiquitous and on-demand access to a shared pool of resources to meet users' requirements [1]. Figure 2.1 depicts an overview of the cloud computing technology.

As depicted in Figure 2.1, cloud computing consists of several shared computer resources, such as computing devices, software, and applications, which are accessed in an on-demand basis. Additionally, different consumers, who access the cloud platform through mobile phones, laptops, desktops, and other peripheral devices, are also present in the cloud computing technological framework. All the resources and consumers are linked through Internet connectivity. The entire framework is part of a cloud platform. Therefore, a cloud platform helps its users to store and process their data in data centers. The important feature of cloud computing is that a central management entity controls all operations associated with different users situated in different geographical regions. Thus, multiple agents can participate in a single platform to share/procure resources in an

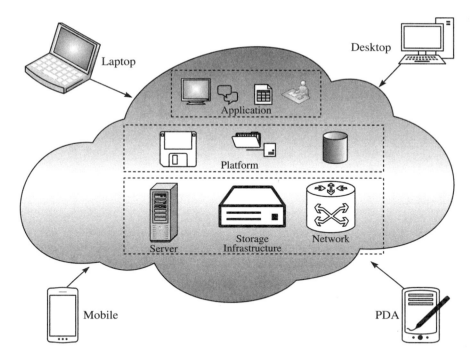

**Figure 2.1** Overview of cloud computing technology

on-demand basis. Figure 2.2 shows an abstracted view of the cloud computing technology. The cloud manager maps the incoming requests with the available resources, which are shared with the cloud manager. So, it is evident that the cloud manager only manages the available resources and incoming requests in an optimum manner, so that the users of the cloud-services can get seamless and ubiquitous services. However, the cloud-service subscribers do not know from which hardware/software resources, the requested services are provided. Similarly, the owners of the shared resources also do not know to whom their resources are mapped. Therefore, although a common platform is used for multiple parties, an abstraction level is always used for security and privacy concerns. As evident from Figure 2.2, there are multiple parties that participate in the resource sharing/procuring/managing process – publisher, subscriber, and manager. The publishers publish their shared resources to the cloud-service providers. On the other hand, subscribers request the desired services to the cloud-service provider. Finally, the cloud-service provider (i.e., cloud manager) manages the available resources shared by the publishers and the incoming requests from the subscribers.

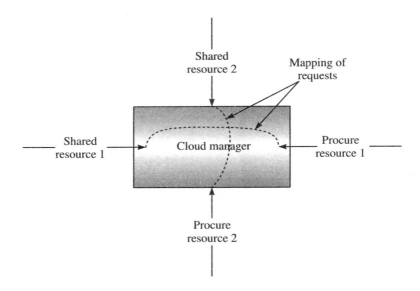

**Figure 2.2** Abstracted view of sharing and procuring services through a cloud platform

## 2.1 Allied Computing Models

Cloud computing adopts different existing architecture, techniques, and tools to offer different computational services. The characteristics of cloud computing have similarities with other computing models, as will be explained in this section.

### 2.1.1 Mainframes

Mainframes are often referred to as computers having the capability of large-scale processing and computation power [2]. They are primarily used by organizations to execute real-time applications, bulk data processing, enterprise resource planning, and banking transactions. The design of a mainframe computer considers the following aspects:

- Redundant engineering for improved reliability and security
- Massive input/output facilities to help in offloading workload to other computers
- Higher utilization of hardware and other computational units in order to provide high throughput

Mainframe computers are introduced for providing large-scale computation facilities required in organizations. Modern mainframe computers are continuously being developed to provide more computation power, while having massive number of sophisticated input and output attachments. The following additional features are included:

- More number of processors with increased clock speed
- Automatic fault detection and self-healing mechanisms
- Increased physical memory and processor core to provide faster computation
- Increased security features
- Support of new hardware and software without service disruption

Mainframes are widely used in automatic teller machines (ATMs), that is, they are widely used to handle banking transactions (order of thousands per second).

The strengths of mainframe – reliability, availability, and serviceability – taken together, are often termed as RAS. Let us discuss these in brief.

**Reliability**: Modern mainframe computers are capable of self-recovery. Therefore, they have extensive self-checking and self-healing capabilities to detect problems. A mainframe system is also capable of offering quick updates to detect problems with the help of new software.

**Availability**: Availability is another important aspect of mainframe computers. A mainframe system provides uninterrupted services to other running systems when there is partial hardware system failure.

**Serviceability**: Finally, once a failure is detected, extensive serviceability of the system allows the replacement of software/hardware with minimum impact on the operating system.

Figure 2.3 shows an example of mainframe workloads for two different scenarios – processing batch jobs and online transactions (adopted from [2]). Further, Figure 2.4 depicts a schematic view of the mainframe world [2], that is, how the interaction between end-users/applications and service providers is done. We limit our discussion on mainframes in this book. However, interested readers are referred to [2], for further information on this topic.

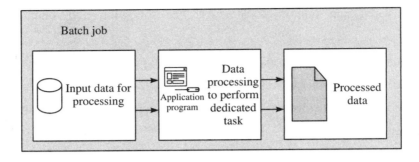

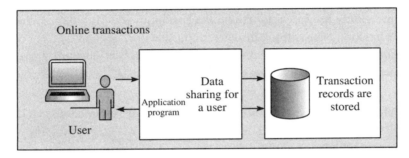

**Figure 2.3** Example of mainframe workloads: Batch job and online transaction

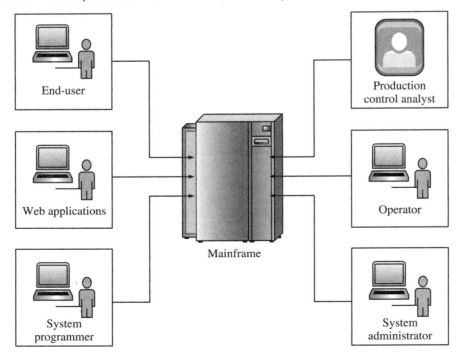

**Figure 2.4** Schematic view of mainframe world

## 2.1.2 Client–server architecture

A client–server architecture has two entities – client and server [3]. The client and the server are connected through a computer network, which can be situated within a system, the same network or different network. A server hosts different services provided by its service providers. On the other hand, a client subscribes for requests over a computer network. Typical examples of client–server services include email, network printing, and websites. Figure 2.5 shows a schematic view of the client–server architecture, in which a server hosts multiple services and clients procure those.

Servers are categorized according to the services they host. For example, a server that hosts web-services is known as a web-server. On the other hand, a server that hosts computer files is known as a file-server. In the same way, there are several types of servers that are categorized according to the services they host. A single server can host multiple services, i.e., a server can host both web-services and file-services. Similarly, a client can also subscribe for multiple services at a time. These are done based on specific application protocols. The most important concern about the client–server model is data synchronization. In other words, synchronization between two updates on specific data is required. Otherwise, we may have inconsistency between applications.

The main principle behind the client–server model is to abstract the underlying technicalities. For example, as a client, a user does not need to be concerned about the associated programs and technical issues to host a file-service and so on. This abstraction is made possible by using application programming interfaces (APIs). APIs hide the complexities of a server from the client. Clients only access the provided services and pay the associated service costs. A client–server architecture is further categorized as a two-tier architecture or a three-tier architecture. Both architectures are briefly discussed here.

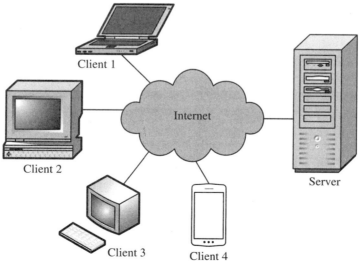

**Figure 2.5** Schematic view of client–server architecture

**Two-tier architecture**: In such an architecture, the client directly accesses the services provided by the server without having an intermediate server. Therefore, the server handles all incoming requests from clients directly. This architecture is typically used in small organizations in which not more than 50 clients are involved. Figure 2.6 shows a schematic view of the two-tier architecture of the client–server model.

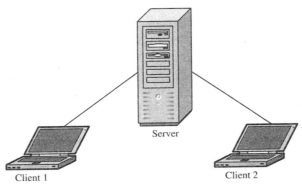

**Figure 2.6** Schematic view of a two-tier client–server architecture

**Three-tier architecture**: In contrast to the two-tier architecture, a middleware is involved in the three-tier architecture. The clients send their requests to the middleware, and the latter manages the requests and send them to the suitable server depending on the service requests. Therefore, in such an architecture, multiple servers co-exist to provide different services to clients. However, the clients are unaware of the presence of such a layering architecture. Figure 2.7 depicts a schematic view of the three-tier client–server model. As discussed before, the cloud computing technology use such architecture, in which the cloud manager arts as the middleware.

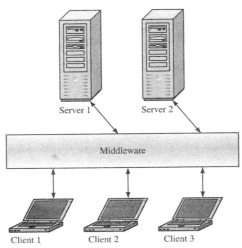

**Figure 2.7** Schematic view of a three-tier client–server architecture

## 2.1.3 Cluster computing

As the name suggests, a group of computers work together to perform some dedicated tasks [4]. There exists a scheduler, which schedules the incoming jobs to pre-designated computers. It is to be noted that all the computers are set to perform the same task. The computers are either loosely or tightly coupled, and typically, they are connected using a high-speed local area network. Further, all the computers (used as servers) have similar hardware configuration and operating system. However, in some specific scenarios, the computers may have different hardware configurations and operating systems. The advantages of cluster computing include high availability and cost-effectiveness compared to those of a single computing unit. Two main aspects – load balancing and high availability – are discussed in brief here.

**Load balancing**: Load balancing is one of the important aspects of cluster computing. The resources of the computers are shared to execute incoming requests. Consequently, a load-balancer assigns the incoming requests to the computers based on computational power. For example, in case of web-service, incoming requests can be assigned to the computers in a *round-robin* manner. On the other hand, jobs with heavy computation task should be assigned to appropriate computers. In such a case, the round-robin based scheme may not give optimal performance. Figure 2.8 shows an example of the cluster computing technique with load balancing.

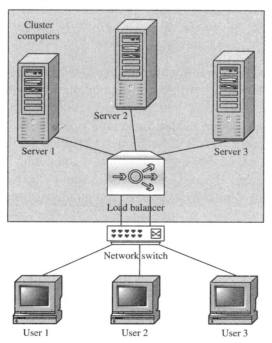

**Figure 2.8** Example of cluster computing with a load-balancer

**High availability**: Another important aspect of cluster computing is high availability. In contrast to the single computing unit-based approach, we have multiple computing units in a cluster. As a result, incoming jobs can be assigned to computers according to their requirements, while utilizing the maximum capacity. On the other hand, in case of a single computing unit, an incoming request may not be served due to insufficient resource availability, although the resources are under-utilized. Another major disadvantage of using a single computing unit is the chance of a single point failure. However, using cluster computing, a single point failure can be handled by assigning the tasks to another computing unit. Therefore, cluster computing provides high availability compared to single computing units.

### 2.1.4 Grid computing

A grid is a collection of multiple computer resources shared from different geographical regions to execute a common goal [5]. In contrast to a cluster, the computational resources may be heterogeneous and distributed in nature in the case of grid computing. The computers in grid computing are responsible for executing different tasks or applications. Therefore, the computers in a grid computing environment may not be available always; they may disappear after executing a particular task. It is to be noted that all the computer resources may not be trustworthy, as they are distributed in nature and are owned by multiple parties. Therefore, the service providers of grid computing should take necessary measures to prevent any misuse of such computing facilities. There exists two core entities in grid computing – the provider side and the user side.

**Provider side**: The provider side is responsible for providing different services through the grid computing platform. Further, the grid computing platform consists of three entities – grid middleware, grid-enabled utility and applications, and software as a service. The grid middleware is responsible for providing a common platform to integrate heterogeneous computer resources together. This is done through virtualization (will be discussed later). Grid-enabled utility and applications provide different computing facility and applications that might interest its users. Finally, software-as-a-service (explained in Section 2.4.3) provides applications that are based on a set of common programming and data structure definitions.

**User side**: The users procure services that are offered by a grid computing platform in an on-demand basis.

Figure 2.9 provides a schematic view of grid computing architecture. Different computer resources are shared with the grid computing management unit in a distributed manner. The administrator manages all the shared resources and incoming requests from users.

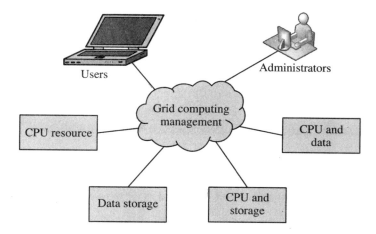

**Figure 2.9** Schematic view of grid computing architecture

## 2.1.5 Service oriented architecture (SOA)

SOA is a special type of software design that provides different services to users through different application protocols [6]. Therefore, a service associated with a user can be accessed and modified independently from other users. Further, a service has few properties, according to [7], as follows:

- Logical representation of business activity is present, while having specific outcomes.
- A service is independent of other services.
- Users are abstracted from the underlying architecture.
- It may contain other underlying services.

Multiple services can be integrated to provide large-scale software application. Figure 2.10 shows an architectural view of SOA.

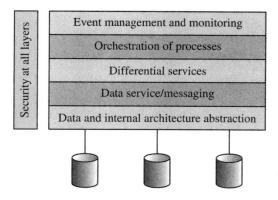

**Figure 2.10** Schematic view of service oriented architecture

All the layers are secured by different security policies; they are also abstracted from the users. An SOA building block can be either a service provider, a service broker or a service requester. The service provider provides the actual services that are requested by a user. On the other hand, a service broker manages the available services offered by the service providers and maps the services to the intended users. Finally, the service requester procures the services according to their requirements. Following are some of the specific technologies that are used to provide independent services to the users:

- Web-service definition language (WSDL) and simple object access protocol (SOAP) are used to provide web-services.
- Messaging is done using message brokers such as *ActiveMQ* and *RabbitMQ*.
- HTTP requests are handled by RESTful application programming interfaces.

### 2.1.6 Utility computing

Utility computing is quite similar to grid computing in the sense that it also offers a service provisioning model that provides different services to the users with maximum utilization of computer resources and minimum associated cost [8]. The users request specific services and the service providers offer the corresponding services on a demand basis. Therefore, in utility computing, we also have the concept of virtualization.

### 2.1.7 Pay-per-use model

The pay-per-use model allows the users to pay only for the services they have procured. In the traditional service provisioning system, the users are always billed according to a static cost, irrespective of their usage. However, in the modern business model, the pay-per-use concept is used to bill the users according to their usage. Such a policy is fair compared to the traditional one.

## 2.2 Virtualization

Virtualization is a technique that is used to efficiently share the computing resources virtually (instead of creating them actually) [9]. However, as a user, one feels the virtual system as an actual one. Cloud computing uses this concept to provide services to its users in an on-demand and pay-per-use basis. Virtualization is categorized as follows – hardware and other virtualization. Hardware virtualization refers to platform virtualization, in which hardware resources of a host machine are shared with a guest machine. As a result, the guest machine looks like a real computer with an operating system. For example, a computer running with a Windows operating system can host a Linux operating system through hardware virtualization. It should be noted that the hardware resource of the host machine is actually shared with the guest. We should not be confused with the term *virtual*. We should keep in mind that the hardware resources always physically exist. It is not possible to create hardware resources that do not exist physically. The software used to enable virtualization of a system is known as hypervisor. Hardware virtualization includes the virtualization of memory, processor, input/output modules, and so on. Figure 2.11 shows an example of hardware virtualization. Other virtualization techniques include software, database, and data virtualization. The hypervisor helps to create virtualized platforms.

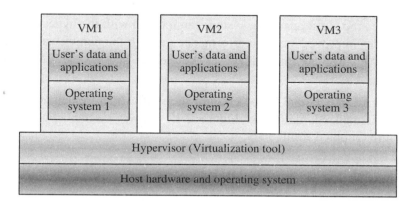

**Figure 2.11** Example of hardware virtualization

## 2.3 Hypervisor

The software which helps to create virtual platforms is known as the hypervisor. Hypervisors are categorized as type 1 and type 2. Type 1 hypervisors run directly on hardware resources without relying on the host operating system. Therefore, they are independent of the host operating system. This type of hypervisors are mainly used to create large-scale virtualized systems, in which the hardware resources are geographically distributed over different locations. On the other hand, type 2 hypervisors rely on the host operating system. They cannot be run without a host operating system. This type of hypervisors are mainly used for desktop virtualization where a host operating system always exists. Let us focus on the hypervisors (type 1) widely used in the cloud computing platform. Some of the notable hypervisors are VMWare Esxi, Xen, and Kernel-based virtual machine (KVM).

- **VMWare Esxi**: This is a type 1 hypervisor used to create virtual machines on a cloud computing platform. According to its type, it does not require a host operating system. It supports live migration of VMs, that is, a VM can be moved from one physical machine to another without disrupting the ongoing services. It allows advanced virtualization techniques through which more number of VMs can be hosted inside a single physical server. However, it has few limitations as follows:

– Infrastructure: It supports the physical resources of host and guest machines upto a maximum limit. It is incapable of creating virtual machines beyond that limit. For example, in a guest VM, it supports maximum 4080 GB RAM and 6 TB RAM in host machines.
– Network: It supports a certain number of virtualized channels and interfaces per host/switch.

- **Xen**: It was started as an open source project; it supports the paravirtualization concept, i.e., the guest operating system can interact with the hypervisor. Therefore, Xen provides improved system performance through the paravirtualization concept. Later on, Xen was commercialized as XenServer and Oracle VM besides being open source.
- **KVM**: It is a Linux kernel-based virtualization infrastructure that acts as hypervisor to create virtualized platforms. Figure 2.12 shows a schematic view of KVM architecture (adopted from [10]).

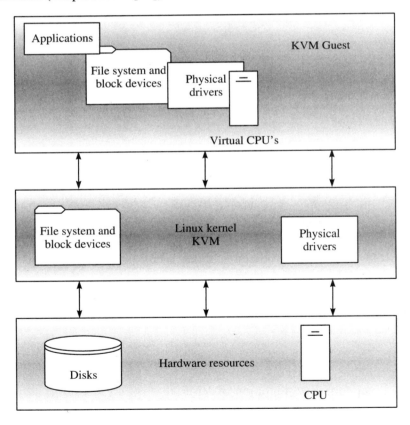

**Figure 2.12** Schematic view of KVM architecture

## 2.4 Types of Services

Typically, cloud computing technology provides three major services – infrastructure as a service (IaaS), platform as a service (PaaS), and software as a service (SaaS) – which are discussed in the section.

### 2.4.1 Infrastructure as a service

Infrastructure as a service (IaaS) is the most basic cloud-service offered by cloud-service providers [11]. Typically, IaaS includes storage, virtual machines, and network support. Therefore, using IaaS policy, a cloud-service provider offers physical infrastructure to its users to execute operations. As discussed earlier, it abstracts the physical properties of the provided infrastructure, such as location, back up policy, security, data partitioning, and detailed computing resources, from its users. Currently, several hypervisors (such as VMWare, Kernel virtual machine, and Virtualbox) are available to create virtual machines as guests using the existing infrastructure resources. However, hypervisors do not reveal the hardware properties of the host machines. The IaaS cloud-service also offers other services such as firewalls, security, virtual network, and software bundles to the virtual machines. Currently, IaaS is the biggest cloud-service offered by the cloud-service providers. All the services are provided to the users in real-time, while pooling the resources available in data centers maintained by the cloud-service providers. Additionally, the users can also share their hardware resources with the cloud-service providers that can be used to host different services. Examples include Amazon web-service (AWS), Microsoft Azure, IBM cloud, and Google cloud. Among the available IaaS platforms, AWS shares the largest platform to provide IaaS to users as of today.

### 2.4.2 Platform as a service

Platform as a service (PaaS) is another important service offered by cloud-service providers after IaaS cloud-services [13]. Through PaaS, development and delivery of different programming models are supplied to the IaaS. Consequently, the users can access the programming models through the cloud platform. Therefore, it can be conceptualized as a computing platform to the cloud users. Different computing platforms include the operating system, multiple languages, and database. Currently, some of the cloud-service providers (such as Microsoft Azure) offer PaaS service, in which the computing resources and storage are automatically scaled up/down depending on application-specific requirements. Therefore, users are not required to select the required resources manually. Further, the PaaS model is categorized into two sub-models – integrated PaaS (iPaaS) and data PaaS (dPaaS). Using the iPaaS model, customers are able to develop and monitor integrated flows within the cloud-service. On the other hand, dPaaS enables cloud-service providers to control their network and execute data within the cloud-service model. Therefore, using the iPaaS model, cloud-services are controlled from the customers' ends. In contrast, using the dPaaS model, cloud-service providers control the execution and networking tools in the cloud platform. As a whole, we can say that PaaS is responsible for executing different programs provided from customers in run-time. Thus, it offers execution of services without downloading or installing the required software.

### 2.4.3 Software as a service

Finally, software as a service (SaaS) is another type of service offered by the cloud-service providers [14]. Using the SaaS model, users access different software and programs, and database. The cloud-service providers manage the required infrastructure and platform to run the required applications. Typically, SaaS is often termed as the 'pay-per-use' model, i.e., the cloud subscription fee is calculated based on the services obtained through the SaaS model. The cloud-service provider installs the required software and programs at the cloud platform, so that the required application can be executed. Thus, the users are not required to install the required software and programs to run the desired applications. As a result, SaaS eliminates the users' burden of installing/buying the required software. It should be noted that cloud applications are different from general applications. Cloud applications can be cloned into multiple virtual machines for parallel execution. However, the cloud users always see that the program is executed at a single point, but internally, it may be distributed over multiple virtual machines. Additionally, a particular virtual machine can serve multiple cloud users.

Figure 2.13 depicts different properties, service models, and deployment models of cloud computing technology. Different properties and deployment models are discussed in the subsequent sections.

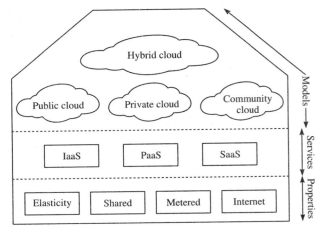

**Figure 2.13** Different services, models, and properties of cloud computing

## 2.5 Types of Deployment

Similar to the service models, cloud computing has several deployment models – public cloud, private cloud, hybrid cloud, and community cloud [12]. Figure 2.14 shows a schematic view of the cloud deployment models. Let us discuss each of the cloud deployment models in detail.

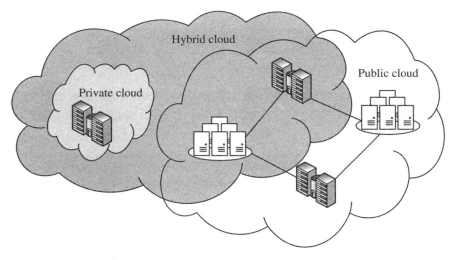

**Figure 2.14**  Schematic view of the cloud deployment models

### 2.5.1  Public cloud

As the name suggests, services provided by a public cloud are open and accessible to all public users. Therefore, such clouds are deployed to help general users in computation, execution, or storage of data. The users can be from different geographical locations, but they must be registered to the cloud-service provider to access different services. For example, Amazon web-service (AWS) [15] is one of the most popular public cloud-service models. In this model, the cloud-service providers maintain their infrastructure at the data center end, and required services are provided to the users through the Internet. Moreover, a cloud-service provider may render services from other cloud-service providers to meet users' requirements.

### 2.5.2  Private cloud

In contrast to the public cloud, private cloud deployment models restrict access from public users. Typically, private clouds are deployed to meet the requirements within an organization. Therefore, it can be accessed within that organization only, while also integrating additional features. For example, if an organization XYZ maintains cloud-service for its employees, the employees of the organization can access the cloud-services from any place over the Internet. Moreover, this access can also be limited within the office premises itself. For example, when the employee is in office, he/she would be able to access the cloud-service. However, outside the office, it cannot be accessible for security reasons. Therefore, the resources are accessible within internet, but not over the Internet. As we can see, organizations need to manage the cloud-service models; they may incur more cost than when obtaining services from public clouds. As a result, organizations need to make a trade-off between the deployment cost and security concerns with private and public cloud models, respectively.

### 2.5.3 Hybrid cloud

Hybrid cloud is an integration of public and private cloud models that capture the advantages of both of the models. In hybrid cloud models, two or more clouds remain distinct in nature, but they are bounded together to offer the benefits of multiple cloud deployments. For example, an organization keeps all secret data (such as company policy) in its private cloud, but shares all other information in a public cloud, so that general customers get access to those information to know more about the company. Moreover, it may happen that some of the services cannot be fulfilled using private cloud, in which case, public cloud models are used within an organization. Therefore, a restriction model is followed within organizations while integrating multiple public and private clouds together.

### 2.5.4 Community cloud

A group of users can maintain a common cloud platform, which is termed as community cloud. It can only be accessed by the users who belong to the community. For example, multiple IT companies may build a community cloud to serve specific purposes that are common to them. It reduces implementation and operational cost for organizations.

## 2.6 Advantages of Cloud Computing

In the previous sections, we discussed the properties and deployment models of cloud computing technology. Let us focus on some of the important features of cloud computing that made it so popular among the users.

### 2.6.1 Elastic nature

The most important feature of cloud computing is its elastic nature. As discussed earlier, it provides flexible service models to the users, i.e., it can expand or reduce the resources to the users depending on application-specific requirements. Therefore, we do not have dedicated hardware/software models to execute one type of application. The same hardware and software resources can be used to execute multiple applications, which benefit the users and help them to avoid incurring higher infrastructural and operational costs.

### 2.6.2 Shared architecture

Another important feature is that it supports a shared architecture, i.e., multiple resources can be shared on a common platform. Depending on the requirements from users, the shared resources can be pooled to fulfill the requirements. Again, cloud computing does not have dedicated resources; they are pooled from multiple sources. The same resources are utilized to serve multiple applications. Moreover, the users think that they access the services from a single entity in a seamless manner. Thus, the cloud manager intelligently maps the requests to the available shared resources, so that the users do not experience any disturbance in the cloud-service.

### 2.6.3 Metering architecture

Cloud computing models also follow the pay-per-use policy, similar to the case of electricity services. In such a model, cloud users pay according to the service they use from the service providers. Therefore, the adoption of cloud computing technology reduces the service cost to the users without the deployment of any dedicated hardware resources at their end. The cloud-service providers keep a record of the services incurred by its users, and automatic billing is done based on the procured services.

### 2.6.4 Supports existing internet services

More importantly, cloud computing supports the existing Internet architecture through which the services can be provided to cloud users. Therefore, there is no need to invest in the Internet services to support cloud-services, unlike other services such as supporting the requirements of internet of things (IoT), in which heterogeneous devices and traffic are expected to be present.

## 2.7 Architecture of Cloud Computing

Figure 2.15 depicts the architecture of cloud computing. As discussed earlier, cloud computing architecture includes cloud-services, platform, storage, and infrastructure. Cloud offers different services that are available within the cloud platform. The cloud platform is responsible for executing the incoming requests from its users. Typically, the cloud storage is maintained by databases. The data is stored and read from the database, which is used later for different purposes, for example, billing the user for procured services. Finally, the cloud infrastructure is responsible for providing the infrastructure necessary to execute the cloud platform tasks. Additionally, it is also responsible for billing the virtual machines from which different services are offered.

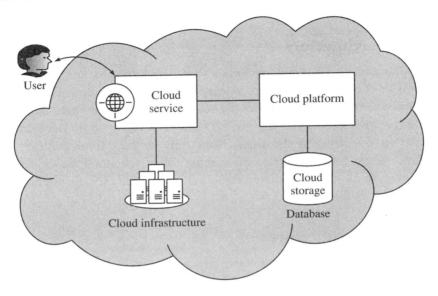

**Figure 2.15**   Cloud computing architecture

## 2.8  Summary

In this chapter, we discussed the cloud computing technology, while explaining different service models, deployment models, and their properties. It is evident that the integration of cloud computing technology is capable of fulfilling different requirements of smart grid in an efficient manner.

### Test Your Understanding

Q01. Define cloud computing.

Q02. State the types of services provided by the cloud computing technology.

Q03. Briefly describe the types of services provided by the cloud computing technology.

Q04. State the deployment models of cloud computing.

Q05. Mention the features of cloud computing.

Q06. Describe the architecture of cloud computing.

# References

[1] Bera, S., S. Misra, and J. J. P. C. Rodrigues. 2015. 'Cloud Computing Applications for Smart Grid'. *IEEE Transactions on Parallel and Distributed Systems* 26 (5): 1477–1494.

[2] Mainframe concepts, IBM. Accessed 30 August 2017. Available at https://www.ibm.com/support/knowledgecenter/zosbasics/com.ibm.zos.zmainframe/zmainframe_book.pdf

[3] Kurose, J. F. and K. W. Ross. 2005. *Computer Networking: A Top-down Approach Featuring the Internet*. India: Pearson Education.

[4] Buyya, R. 1999. Accessed 30 August 2017. Available at http://www.buyya.com/cluster/.

[5] Berman, F., G. Fox, and A. J. G. Hey. eds. 2003. *Grid Computing: Making The Global Infrastructure a Reality*. New York: John Wiley & Sons.

[6] Erl, T. 2005. *Service-Oriented Architecture Concepts, Technology, and Design*. New Jersey: Prentice Hall.

[7] The Open Group. Accessed 15 September 2017. Available at http://opengroup.org/standards/soa.

[8] Garfinkel, S. 1999. *Architects of the Information Society: 35 Years of the Laboratory for Computer Science at MIT*. Cambridge, MA: MIT Press.

[9] Buyya, R., J. Broberg, and A. M. Goscinski. 2010. *Cloud Computing: Principles and Paradigms*. New York: John Wiley & Sons.

[10] Huynh, K. and S. Hajnoczi. 2010. 'KVM / QEMU Storage Stack Performance Discussion'. Paper presented at the Linux Plumbers Conference.

[11] Serrano, N., G. Gallardo, and J. Hernantes. 2015. 'Infrastructure as a Service and Cloud Technologies'. *IEEE Software* 32 (2): 30–36.

[12] Mell, P. and T. Grance. 2011. *The NIST Definition of Cloud Computing*. US: US National Institute of Science and Techonology Std. Accessed 16 September 2017. Available at http://csrc.nist.gov/publications/nistpubs/800-145/SP800-145.pdf.

[13] Dikaiakos, M.D., D. Katsaros, P. Mehra, G. Pallis, and A. Vakali. 2009. 'Cloud Computing: Distributed Internet Computing for IT and Scientific Research'. *IEEE Internet Computing* 13 (5).

[14] Armbrust, M. A., Fox, R. Griffith, A. D. Joseph, R. Katz, A. Konwinski, G. Lee, D. Patterson, A. Rabkin, and M. Zaharia. 2010. 'A View of Cloud Computing'. *Communications of the ACM* 53 (4): 50–58.

[15] Miller, F. P., A. F. Vandome, and J. McBrewster. 2010. *Amazon Web Services*. London: Alpha Press.

# CHAPTER 3

# Introduction to Big Data Analytics

Big data is conceptualized as a large data set with high complexity that cannot be managed with the traditional data processing schemes in an adequate manner. Big data technology is described as a holistic information management of new types of data alongside traditional data [1]. For example, traditionally, the backbone network carries different homogeneous information generated by end-users. In addition to this, due to the deployment of more number of sensors, actuators, and other different data sources, the network needs to carry heterogeneous information. It is also expected that the same network channel would be used to carry this heterogeneous information. The challenges in big data analytics include analyzing, capturing, processing, searching, storing, and visualizing the large data set. The term 'big data' refers to predictive analysis, i.e., predicting the future condition based on the results obtained from processing existing data. Thus, we can predict the system status in advance to take adequate measures to deal with situations in real-time.

According to the existing literature, there is no hard and fast definition of big data. As mentioned earlier, big data is a conceptualized term that is used to represent a large data set, which is complex and heterogeneous in nature. Moreover, it is hard to process such large data set using the traditional applications. Due to the growing need to digitize data, it is required to store and process large amounts of data generated from heterogeneous devices. To get an idea about the amount of data generated over several past years, Figure 3.1 presents certain statistics regarding growth of information storage capacity with digitization of data (adopted from [2]). As can be seen in Figure 3.1, there is a large amount of data being stored in digital storage devices. However, we need efficient processing applications to process such large amount of data before storing it for future use. Different attributes of big data are discussed in the subsequent sections.

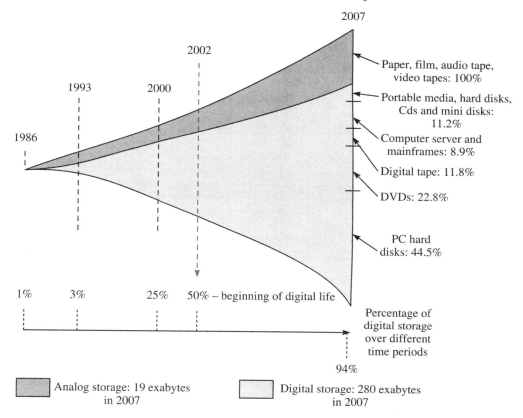

**Figure 3.1** Growth of information storage capacity with digitization

## 3.1 Attributes of Big Data

As the name suggests, one of the main attributes of 'big data' is *volume*, i.e., the size of the data. However, it is not only about the volume, but also other factors such as velocity and variety of the data being analyzed. Figure 3.2 presents the attributes diagrammatically. Let us discuss the attributes in detail.

### 3.1.1 Volume of data

The primary attribute of big data is its volume. Presently, organizations manage a few terabytes or even petabytes of data, which is also stored for processing and analyzing. Further, the volume attribute is quantified in many respects such as size, number of records, transactions, and number of tables and files. A person can think about the volume

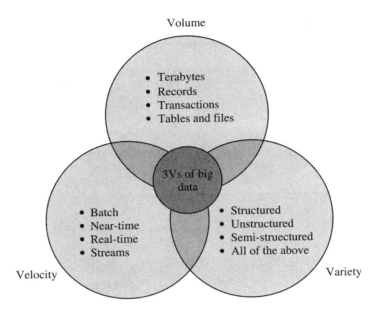

**Figure 3.2**  Different attributes of big data

of big data in terms of storage required, whether it is in the form of gigabytes, terabytes, or petabytes. Most of the time, terabyte is used to represent big data. On the other hand, number of records can also be used to measure big data. For example, in a database, billions of records are stored, so it can be termed as big data in terms of number of records. Concurrently, the number of transactions is also considered as part of big data attributes. Finally, the number of tables and files maintained/stored in database/storage devices are also considered to quantify big data.

### 3.1.2 Velocity of data

Velocity is another important attribute to define big data. It captures the data speed, i.e., how frequently data are generated and sent through the communication network to the data center. For example, data-streaming can be used to define the velocity of data. Velocity also consists of multiple sub-attributes such as batch, near-time, real-time, and streams. A large dataset can be sent through the communication network, spanning multiple applications with all the applications bundled as a batch file. Therefore, although the speed of the entire data set is relatively slow, the receiver thinks that the velocity is high as it continuously receives data for a long time. On the other hand, some data are sent in near-time to the data center. For example, day-ahead energy consumption data by the smart meters can be sent at the end of the day to utility providers. Such data received by the utility providers has greater velocity and volume since there are more than billions of smart meters sending their energy consumption information at the end of the day. Concurrently, real-time data is also an important factor. Presently, there is an interest in Internet of things (IoT), in which multiple sensors and actuators send their sensed and actuated data, respectively, to the backbone networks. The network is also required to send such data in real-time to take adequate decisions.

### 3.1.3 Variety of data

Variety is another important attribute that is used to define 'big data'. Typically, variety of data suggests that the data is generated from different sources that are heterogeneous in nature. As discussed earlier, an IoT environment consists of heterogeneous sensors and actuators to sense and actuate real-time data, respectively. Therefore, the data generated from such heterogeneous devices are also heterogeneous in nature and need adequate care for efficient storage and processing. Additionally, data from Web logs, click-streams, and social media are important factors to consider. It may be argued that the service providers maintain such Web logs, records, and data generated from social media. However, the growing interest in the digitization of everything poses challenges to researchers and cause them to rethink data handling techniques. Different varieties of data include structured, unstructured, and semi-structured data, which are required to be stored and processed in an efficient manner. In case of structured, the heterogeneous data obtained from different sources can be mapped into a pre-defined format. For example, storing of incoming data in an SQL database. Therefore, structured data always has a specific format, which makes it easy to store them in a database. In case of semi-structured data, the incoming data is somewhat organized, but requires some changes before it can be stored in the database. For example, the data coming in JSON format (widely used in sensor network) need to be analyzed before storing. In contrast to the structured and semi-structured data, unstructured data do not have any specific property or relational key. This type of data cannot be fitted directly into the database. In practical scenarios, around 80% data is unstructured data. Example includes satellite data, scientific data, photographs, sonar or radar data, and so on.

## 3.2 Overview of Big Data Analytics

As discussed in Section 3.1, big data analytics refers to the collection of data from various sources and conversion into useful format for organizational business. According to Cross Industry Standard Process-data mining (CRISP-DM) [3], the life cycle of data consists of six stages – business understanding, data understanding, data preparation, modeling, evaluation, and deployment. Figure 3.3 shows the relationship among all these stages (adopted from [3]). Let us focus on each of the stages in detail.

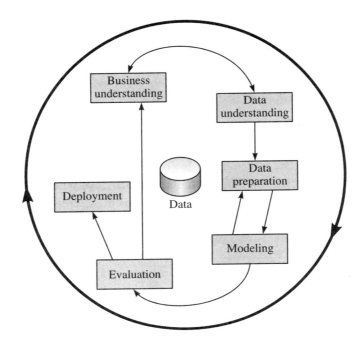

**Figure 3.3** Life cycle of data defined by CRISP

- Business understanding: This stage focuses on understanding business objectives. According to the business model, the initial plan is created to achieve the objectives. Then, a decision model is prepared to take further actions. This stage includes the following sub-stages:

    – Determine business objectives
    – Assess situation
    – Determine data mining goals
    – Produce, project, and plan

- Data understanding: This stage starts with initial data collection. Concurrent to the data collection, different activities are carried out to get familiar with the data. Further, subsets of data are analyzed according to the business understanding. This includes the following sub-stages:

    – Collect initial data
    – Describe data
    – Explore data
    – Verify data quality

- Data preparation: In this stage, the final data set is prepared from the collected data. Data preparation tasks include preparing the data in tabular format, recording of data, attributes selection, and transformation of data according to the tools used in the modeling stage. This also includes a few sub-stages, as follows:

- Select data
- Clean data
- Construct data
- Integrate data
- Format data

- Modeling: In this stage, data calibration is done on the prepared data by using different models in order to have optimized dataset according to the business requirement. Further, the modeled data is used to understand the incoming data from various sources, as mentioned earlier. The sub-stages included in this stage are as follows:

  - Select modeling techniques
  - Generate test design
  - Build model
  - Assess the model

- Evaluation: In this stage, the models are evaluated based on the obtained results. Consequently, the best model is chosen for deployment in the next stage. It includes the following sub-stages:

  - Evaluate results
  - Review process
  - Determine next steps

- Deployment: Finally, deployment of the models is required to generate report. The report should be simple to understand. The generated report helps to understand the customers' preferences and choices. This stage includes the following sub-stages:

  - Plan deployment
  - Plan monitoring and maintenance
  - Produce final report
  - Review project

However, big data analytics cycle is quite different from the CRISP-DM model. The big data analytics includes the following stages [4]:

- Business problem definition: This stage is similar to the business understanding stage defined by CRISP. In this stage, the business value is evaluated based on the cost and expected outcome.

- Research: Methodologies that are appropriate for the business strategy are evaluated and used in future stages.
- Human resources assessment: This stage involves analyzing the manpower – whether current manpower is capable of handling the project successfully.
- Data acquisition: This stage focuses on the data collection from various sources.
- Data munging: After data collection, the collected data are required to be stored in a precise format for further processing. In this stage, the format of the data is defined before storing the data.
- Data storage: In this stage, the retrieved data is stored according to the format defined in the data munging stage.
- Data analysis: The stored data is understood before applying the modeling tool.
- Data preparation for modeling: The stored data is further prepared for modeling in this stage.
- Modeling: Different modeling tools are applied on the prepared data in order to get optimal results according to the business requirements.
- Implementation/deployment: Finally, a suitable model is employed to analyze the data, which helps in determining customers' preferences and market status. This is similar to the deployment stage defined by CRISP.

## 3.3 Benefits of Big Data Analytics

When the received (big) data is analyzed to make predictions or to take decisions, it is termed as big data analytics. The term 'analytics' refers to the processing of data in an efficient manner and the retrieving of useful output for taking better decisions. Some of the benefits of big data analytics are discussed here.

- Better customer relation: It is expected that big data analytics will improve customer relations with organizations. For example, the organization can analyze the products or features in which the customers are interested. This can be obtained after analyzing multiple responses received from a variety of customers. It includes social marketing, segmentation of customer base, and recognition of market opportunities.
- Improving business intelligence: Business intelligence can be improved using big data analytics. It includes improved planning and forecasting, changing products, and changing business policies. Using big data intelligence, the root causes of high operational cost can also be analyzed.
- Supporting different applications: Some of the specific applications such as satellite images, radar data, and data generated from heterogeneous sources, can also be analyzed using big data analytics, which is expected to help the service

providers/organizations to take better decisions. For example, fraud detection and trending things in marketplace can be predetermined using such analytics method.

Figure 3.4 summarizes the benefits of big data analytics, which includes good policy for customers, better planning, better security, real-time automated processing, etc.

- Social marketing
- Business insights
- Segmentation of customer base
- Sales and marketing opportunities
- Automated real-time processing
- Improved policy for customers
- Segmentation of customer base
- Better security
- Better planning and forecasting
- Others

**Figure 3.4** Benefits of big data analytics

## 3.4 Big Data Analytics for Smart Grid

As discussed in Chapter 1, smart grid consists of several entities that generate huge amounts of data. These data are required to be stored and processed in an efficient manner while considering the energy consumption at the data centers. Smart grid fulfills all the 5Vs of big data – volume, velocity, variety, varacity, and value. In a report, it is mentioned that the utility companies had more than 194 petabytes of data in 2009 [5]. Big data can help in effective management of smart grid in terms of data management from smart meters, billing, phasor management units (PMUs), and management of intelligent devices in the smart grid. Further, big data analytics can help in real-time energy management, demand side management, outage detection, and asset management in the smart grid.

Figure 3.5 presents an overview of the benefits of big data analytics from the perspective of smart grid (adopted from [6]). As mentioned in Figure 3.5, big data analytics can be applied to perform multiple sub-analytics in the smart grid system. For example, from the grid's perspective, weather forecasting is an important feature to predict the impact of renewable energy sources. To be more specific, bad weather can disrupt both solar energy and wind energy generation. On the other hand, big data analytics can also be applied for customers' data analytics. For example, typically, a customer's energy consumption pattern throughout a day is mostly similar to other days. Therefore, energy consumption profiling can help in demand forecasting in advance for improved energy management in the smart grid. Similarly, other aspects can also be accounted for using big data analytics and other data management approaches which are discussed in Part III of this book.

**Figure 3.5** Benefits of big data analytics in the smart grid

## 3.5 Big Data Analytics Tools

Following are some big data analytics tools that are used widely to process large set of data received from multiple sources.

- R (https://www.r-project.org/)
- RapidMiner (https://rapidminer.com/)

- SQL (https://www.mysql.com/)
- Python (https://www.python.org/)
- Excel (https://products.office.com/en-in/excel)
- KNIME (https://www.knime.com/)
- Hadoop (http://hadoop.apache.org/)
- Tableau (https://www.tableau.com/)
- SAS base (https://www.sas.com/en_in/software/base-sas.html)
- Spark (https://spark.apache.org/)

We limit our discussion on the aforementioned analytics tools. However, interested readers can access the given links for further reading.

## 3.6 Summary

In this chapter, we discussed the concept of big data and its different attributes. The attributes include volume, velocity, and variety of data obtained from multiple devices. The advantages of big data analytics were discussed. Due to the inherent features of big data analytics, different analytical methods can be applied to fulfill smart grid information management. Thus, improved smart grid energy management can be performed using the power of big data analytics, which will be discussed later in this book.

---

**Test Your Understanding**

Q01. Define the term 'big data'?

Q02. Describe the different attributes of big data.

Q03. What are the advantages of big data analytics?

---

# References

[1] Oracle. 'Introducing Big Data Cloud at Customer'. Accessed 25 September 2017. Available at https://www.oracle.com/big-data/index.html.

[2] Hilbert, M. and P. Lopez. 2011. 'The World's Technological Capacity to Store, Communicate, and Compute Information'. *Science* 332 (6025): 60–65.

[3] Shearer, C. 2000. 'The CRISP-DM Model: The New Blueprint for Data Mining'. *Journal of Data Warehousing* 5 (4): 13–22.

[4] Tutorialspoint. 2017. 'Big Data Analytics–Basics'. Accessed 5 November 2017. Available at https://www.tutorialspoint.com/big_data_analytics/.

[5] McMahon, J. 2013. 'Utilities Dumbstruck by Big Data from Smarter Grid'. *Forbes.* Accessed 26 September 2017. Available at http://www.forbes.com/sites/jeffmcmahon/2013/09/25/utilities-dumbstruck-by-big-data-fromsmarter grid/.

[6] Asad, Z., M. A. R. Chaudhry. 2017. 'A Two-Way Street: Green Big Data Processing for a Greener Smart Grid'. *IEEE Systems Journal* 11 (2): 784–795.

# CHAPTER 4

# Fundamental Mathematical Prerequisites

In this chapter, some existing mathematical methodologies are discussed that have the potential to solve different problems presented in the subsequent chapters.

## 4.1 Linear Programming

Linear programming can be used for the maximization or minimization of some function while considering some constraints. For example, if we want to travel the maximum distance and we have some money, different travel options (such as bus, private car, train, and shared car) with unit distance cost can be retrieved. Hence, our objective could be to choose a suitable travel option, so that we can travel the maximum distance with the given money. The constraints may have upper and lower bounds.

Mathematically, we can present a linear optimization problem, as follows [1]:

$$\text{Maximize} \quad y_1 + y_2$$
$$\text{subject to} \quad y_1 \geq 0, \; y_2 \geq 0$$
$$y_1 + 2y_2 \leq 4$$
$$4y_1 + 2y_2 \leq 12$$
$$-y_1 + y_2 \leq 1$$

In the aforementioned optimization problem, we have an objective function, which is the maximization of $y_1 + y_2$. We have five constraints. The first two are $y_1 \geq 0$ and $y_2 \geq 0$, known as the non-negativity property. This property can often be found in linear programming. The other three constraints are used to get a bounded region in a 2-D plane, as there are two unknowns (i.e., $y_1$ and $y_2$) in the optimization problem. We can get the values for $y_1$ and $y_2$ easily by plotting the constraints in the *xy*-plane. However, all linear optimization problems cannot be solved so easily due to several facts – for example, one constraint may have an equality while another may have an inequality. There are several approaches to solve linear optimization problems – the Simplex method [2] and online tools (IBM CPLEX) [3] are two popular ones to name a few.

## 4.2 Integer Linear Programming

Integer linear programming (ILP) is an extension of linear programming; there are some additional constraints that are integers. In a pure ILP, all the variables are integers. For example, an organization wants to undertake projects in a financial year. Now, each project has an execution cost and associated revenue. The organization has a fixed budget for the financial year. In such a scenario, some projects need to be selected from many due to budget constraints. Mathematically, the problem is represented as follows:

$$\text{Maximize} \quad \sum_{i=1}^{n} R_i X_i$$

$$\text{subject to} \quad \sum_{i=1}^{n} C_i X_i \leq B$$

$$0 \leq X_i \leq 1, \text{ where } X_i \text{ is an integer}$$

and $R_i$ and $C_i$ are the revenue and execution costs associated with project $i$, respectively. $X_i$ is either 0 or 1, i.e., the project $i$ is either rejected or selected, and $B$ is the budget allotted for the current financial year. Therefore, the overall objective of the organization is to maximize its revenue by executing as many projects as possible within the given budget in the financial year. To solve such types of ILP, the cutting plane algorithm can be applied [4]. The basic steps of the algorithm are as follows:

- Step 1: Solve the problem as a linear programming model by ignoring the integrality constraints.
- Step 2: If the variables are all integers, then an optimal solution can be found.
- Step 3: Otherwise, generate a CUT, it refers to the cutting-plane method that iteratively finds a feasible set or objective function to satisfy the linear inequalities, termed a CUT. i.e., add a constraint that is satisfied by all integer solutions, but not by the current linear programming solution.
- Step 4: Apply the new constraint to the problem, and solve it from Step 1.

## 4.3 Mixed Integer Linear Programming

Similar to the ILP, mixed integer linear programming (MILP) contains integer variables. However, it is not necessary that all the variables are integers. For example, in an electricity distribution system, the objective of a service provider is to minimize electricity production cost from its power plants, while fulfilling the customers' demands. Mathematically, it can be formulated as follows:

$$\text{Minimize} \quad \sum_{i=1}^{n} C_i X_i + C_{\text{const}}$$

$$\text{subject to} \quad \sum_{i=1}^{n} P_i \geq \sum_{j=1}^{k} D_j$$

$$0 \leq X_i \leq 1, \text{ where } X_i \text{ is an integer}$$

and $C_i$ and $P_i$ denote the production cost and electricity produced from power plant $i$. There are two constraints – energy demand $D_j$ from customer $j$ and the status of a power plant (which can be ON or OFF), and is determined by $X_i$. The demand may be non-integer. The total produced electricity must be greater than or equal to the total energy demand from the customers. For simplicity, other constraints such as transmission loss, power plant up-time, and down-time may be relaxed. The solution approach is similar to the solution approach to the ILP problems. However, in many cases, getting an optimal solution to these problems is NP-hard, i.e., an optimal solution cannot be obtained in polynomial time. Therefore, we apply approximation algorithms to solve such problems to get *near* optimal solution, which is discussed later.

## 4.4 Non-Linear Programming

In case of linear programming, the impact of the variables on the objective function is linear. Therefore, we can present a maximization problem $f(x)$ as a minimization problem by representing it as $-f(x)$. However, there are several practical situations in which the objective function is not linearly dependent on its variables. For example, in a smart grid energy trading system, a customer can consume energy from two service providers $S_1$ and $S_2$. Let both the service providers have unit energy costs $p_1$ and $p_2$, respectively. Moreover, let $p_1$ be higher than $p_2$. If the percentages of load-shedding for both the service providers are $l_1$ and $l_2$, respectively, where $l_1 < l_2$, clearly, the customer can consume energy from $S_1$ with higher cost, but in a more reliable manner (i.e., percentage of load-shedding is less). On the other hand, the customer can consume energy from $S_2$ with lower cost while compromising the reliability of energy service. It can be seen as a linear problem, if we fix the other part, i.e., it becomes a minimization of cost problem by putting an upper limit on the load-shedding percentage or a minimization of load-shedding percentage problem by putting an upper bound on the energy consumption cost. However, in a practical scenario, a customer will expect to have a trade-off between the energy consumption cost and the associated reliability[1]. In such a case, the problem cannot be presented/solved using the linear optimization methods, as they do not follow the property of linear programming models. Therefore, we need non-linear programming models to deal with such scenarios. In the smart grid energy management system, we may observe that most of the problems are non-linear in nature. Mathematically, such problems can be presented as follows:

Maximize $\quad \theta(3x_1^2 + x_2^2 + (x_1+x_2)^2) - \alpha x_1 - \beta x_2$

subject to $\quad g_1(x) = x_1 + x_2 \leq 4$

$\quad x_1 \geq 0$, and $x_2 \geq 0$

where $\alpha x_1$ and $\beta x_2$ denote the consumed energy from service providers $S_1$ and $S_2$, respectively. $(3x_1^2 + x_2^2 + (x_1+x_2)^2)$ is the variance of the percentage of load-shedding.

In general, non-linear models are either concave or convex, although some of them are neither concave nor convex. In the concave non-linear model, the objective function is always non-increasing, and in the case of a convex model, it is always non-decreasing. Linear programming models are not directly solved. There are several ways to solve the problems:

- They can be converted to an *unconstrained optimization* problem, i.e., all the constraints in the optimization problem are relaxed. Then, it can be solved easily.
- If the optimization function is quadratic in nature, the problem can be solved using the quadratic programming method.
- If the problem can be divided into two single optimization problems, they can be solved separately using separable programming.
- The non-linear optimization problem can also be solved using convex programming. The property of convexity makes it easier to find out the global minimum instead of the local minimum in an optimization problem. The convexity of an optimization function can be proved through first and second order derivatives.

## 4.5 Quadratic Function

Another important method is through quadratic functions, which are widely used to define the cost function in a smart grid. When we have a function in the form of $ax^2 + bx + c$, where $a \neq 0$, $b$ and $c$ are constants, then the function is called a quadratic function. The curve for a quadratic function is a parabola. If $a$ is positive, the parabola opens upward,

---
[1] Reliability is more when percentage of load-shedding is less and vice-versa.

and if $a$ is negative, the parabola opens downward. We will see later on that this quadratic function is used to determine energy consumption cost in many cases where $x$ is used as energy supply or demand to the grid.

## 4.6 Different Distributions

### 4.6.1 Normal distribution

Normal distribution is a standard distribution technique to represent several natural phenomenon. It has the following simple characteristics:

- It is symmetric in nature, i.e., bell-shaped.
- It is continuous for any value of $X$ between $-\infty$ and $+\infty$. Therefore, the probability of $X$ for any number between $-\infty$ and $+\infty$ has a value other than zero.
- Two parameters – mean ($\mu$) and standard deviation ($\sigma$) – determine the property of the distribution.
- The normal density function is represented as $f(x;\mu,\sigma^2) = \frac{1}{\sqrt{2\pi\sigma^2}e^{-(x-\mu)^2/2\sigma^2}}$.
- The probability of a number lying within one standard deviation of the mean is 0.6826, i.e., $P(\mu - \sigma \leq X \leq \mu + \sigma) = 0.6826$.
- The probability of a number lying within two standard deviations of the mean is 0.9544, i.e., $P(\mu - 2\sigma \leq X \leq \mu + 2\sigma) = 0.9544$.

Figure 4.1 shows a normal distribution over $X$. For example, in a smart grid, energy consumption reading from smart meters can be reported in a fixed time interval. Therefore, such data traffic from smart meters can be considered as a normal distribution.

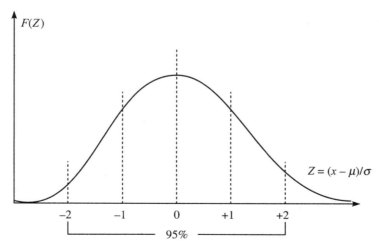

**Figure 4.1** Simplified view of a normal distribution

### 4.6.2 Poisson distribution

In contrast to the normal distribution, the Poisson distribution follows discrete probability distribution of events. Therefore, the number of events that occur in a time interval is independent from other intervals. The probability of $x$ events occurring in a given time interval is represented as follows:

$$P(X = x) = e^{-\lambda} \frac{\lambda^x}{x!} \qquad (4.1)$$

where $\lambda$ is the mean number of events per interval. It is to be noted that a Poisson random variable can have any positive integer value. For example, energy demand requests from customers can be modeled as a Poisson distribution, as any number of customers can send an energy demand request to the grid in a given interval. Moreover, energy demand requests in a particular time interval are independent from other time intervals. In general, incoming traffic in a communication network is modeled as a Poisson distribution.

### 4.6.3 Gaussian distribution

Gaussian distribution is also referred to as a normal distribution with $d$-dimensional Gaussian probability distribution as:

$$P(\mathbf{x}) = \frac{1}{(2\pi)^{d/2} |\Sigma|^{1/2}} \exp\left(-\frac{(\mathbf{x}-\mu)^T \Sigma^{-1}(\mathbf{x}-\mu)}{2}\right) \qquad (4.2)$$

where $\mathbf{x}$ is a $d$-dimensional column vector and $\mu$ is the mean vector. $\Sigma$ is the $d \times d$ covariance matrix. Mathematically,

$$\mu = E[\mathbf{x}] = \int \mathbf{x} p(\mathbf{x}) d\mathbf{x} \qquad (4.3)$$

and

$$\Sigma = E[(\mathbf{x}-\mu)(\mathbf{x}-\mu^T)] = \int (\mathbf{x}-\mu)(\mathbf{x}-\mu^T) p(\mathbf{x}) d\mathbf{x} \qquad (4.4)$$

Gaussian distribution also follows the same property – integral of any probability distribution function with lower limit $-\infty$ and upper limit $+\infty$ is 1. Mathematically,

$$\int_{-\infty}^{+\infty} f(x) dx = 1 \qquad (4.5)$$

where $f(x)$ is a 1-D Gaussian function.

## 4.7 Dimension Reduction Methods

In modeling many systems, the number of variables is very large; multiple sets are also present. Therefore, it may be difficult to consider all the variables and sets directly to model the system. To address this issue, it may be advantageous to consider a linear combination of the variables and sets with some properties such as correlation, covariance, or variance. The selection criteria for such a linear combination of variables and sets depends on the application. The process of finding such a linear combination (with reduced number of variables and sets) from a large number is known as dimension reduction methods. There exists a few well-known dimension reduction methods, which are discussed here.

### 4.7.1 Principal component regression (PCR) method

Principal component regression (PCR) is useful to analyze multiple regression data that suffers from multi-collinearity. In the presence of multi-collinearity in data, we can apply least-square methods to analyze the data. However, the variance can be very large as well, which, in turn, means that the *true* value can be far from the estimated one. To reduce the standard error, a degree of bias can be added to the regression estimate. Thus, principal component regression reduces the standard error.

Let the regression equation be written in a matrix form as follows [5]:

$$\mathbf{Y} = \mathbf{XB} + \mathbf{e} \qquad (4.6)$$

where **Y** is the dependent variable, and **X** is the independent variable. **B** denotes the regression coefficients to be estimated using the PCR method, and **e** the errors or residuals. In brief, the PCR method is outlined as follows.

- Perform principal component analysis (PCA) of the given matrix **X**.
- Store the principal components in vector form **Z**.
- Fit the regression of **Y** on **Z** to obtain least-square estimates of **A**.
- Set the last element of **A** to zero.
- Compute the original coefficients using $\mathbf{B} = \mathbf{PA}$.

### 4.7.2 Reduced rank regression (RRR) method

Reduced rank regression (RRR) is an estimation method in multi-variate regression with reduced rank restriction on the coefficient matrix. Let us focus on the RRR model [7].

First, let us consider the multivariate regression of $Y$ on $X$ and $Z$ of dimension $p$, $q$, and $k$, respectively. Then, $Y_t = \Pi X_t + \Gamma Z_t + \varepsilon_t$, where $t = 1, 2, \ldots, T$, and $\Pi$ and $\Gamma$ are coefficients. So the $\Pi$ has reduced rank less than or equal to $r$, and it is expressed as

$\Pi = \alpha\beta'$, where $\alpha = p \times r$ and $\beta = q \times r$, and $r < \min(p,q)$. Therefore, we have the reduced rank model as follows:

$$Y_t = \alpha\beta' X_t + \Gamma Z_t + \varepsilon_t, \text{ where } t = 1, 2, \ldots, T$$

After modeling the system, let us look into the RRR algorithm denoted as $RRR(Y,X|Z)$. A notation is used for product moments – $S_{yx} = T^{-1} \sum_{t=1}^{T} Y_t X_t'$, and $S_{yx.z} = S_{yx} - S_{yz} S_{zz}^{-1} S_{zx}$, and so on. The steps of the algorithm are as follows:

- Regress $Y$ and $X$ on $Z$, and form the residuals $(Y|Z)_t = Y_t - S_{yz} S_{zz}^{-1} Z_t$, $(X|Z)_t = X_t - S_{xz} S_{zz}^{-1} Z_t$, and the product moments as

$$S_{yz.x} = T^{-1} \sum_{t=1}^{T} (Y|Z)_t (X|Z)_t' = S_{yx} - S_{yz} S_{zz}^{-1} S_{zx}$$

- Solve the eigenvalue problem $|\lambda S_{xx.z} - S_{xy.z} S_{yy.z}^{-1} S_{yx.z}| = 0$, where $|\cdot|$ denotes the determinant. Therefore, the ordered eigenvalues are $\Lambda = \text{diag}(\lambda_1, \lambda_2, \ldots, \lambda_a)$, and the eigenvectors are $V = (v_1, v_2, \ldots, v_a)$, so that $S_{xx.z} V \Lambda = S_{xy.z} S_{yy.z}^{-1} S_{yx.z} V$, and $V$ is normalized, such that $V' S_{xx.z} V = I_p$ and $V' S_{yx.z} S_{xx.z}^{-1} V = \Lambda$. This is known as the singular value decomposition method [8].

- Define the estimators $(v_1, v_2, \ldots, v_r)$ together with $\hat{\alpha} = S_{yx.z} \hat{\beta}$ and $\hat{\Omega} = S_{yy.z} - S_{yx.z} \hat{\beta} (\hat{\beta} S_{x.z} \hat{\beta})^{-1} \hat{\beta}' S_{xy.z}$. Similarly, $\hat{\alpha}$ and $\hat{\Gamma}$ are determined by regression once we have $\hat{\beta}$.

## 4.8 Approximation Algorithms

In many optimization problems, finding an *optimal* solution is NP-hard, i.e., it is not possible to find an *optimal* solution in polynomial time. In such cases, we can have *near optimal* solutions to the problems that can be obtained in polynomial time. The algorithms that can be applied to find such *near optimal* solutions are known as approximation algorithms. There exists the concept of $\alpha$-approximation, in which, an algorithm can be called an $\alpha$-approximation algorithm if and only if the algorithm can find a solution for every instance of the problem. In a minimization problem, $\alpha$ is greater than 1, i.e., the solution found by the algorithm is at most $\alpha$ times the optimum solution. Similarly, for a maximization problem, $\alpha$ is less than 1, i.e., the solution found by the algorithm is at least $\alpha$ times the optimum solution. For example, some of the problems that can be solved using the approximation algorithm are minimum vertex cover, minimum weight vertex cover, job scheduling, and non-uniform job scheduling. For brevity, we limit our discussion on this topic. Interested readers can refer to [6].

## 4.9 Summary

There exists several mathematical methods that are useful to solve different problems involving optimization. In this chapter, we discussed some mathematical prerequisites that are useful to understand complex smart grids. Interested readers can refer to existing books [1, 6], which discuss such mathematical approaches in detail.

# References

[1] Ferguson, R. O. and L. F. Sargent. 1958. *Linear Programming: Fundamentals and Applications.* New York: McGraw-Hill.

[2] Nelder, J. A. and R. Mead. 1965. 'A Simplex Method for Function Minimization'. *Computer Journal* 7 (4): 308–313.

[3] International Business Machines Corporation. 2009. *IBM ILOG CPLEX V12.1: Users Manual for CPLEX.* New York: IBM.

[4] Balas, E., S. Ceria, and G. Cornuejols. 1993. 'A Lift-and-Project Cutting Plane Algorithm for Mixed 01 Programs'. *Mathematical Programming* 58 (1–3): 295–324.

[5] NCSS Statistical Software. 'Principal Components Regression'. Accessed 31 July 2017. Available at http://ncss.wpengine.netdna-cdn.com

[6] Williamson, D. P. and D. B. Shmoys. 2011. *The Design of Approximation Algorithms.* Cambridge: Cambridge University Press.

[7] Johansen, S. 'Reduced Rank Regression'. Accessed 31 July 2017. Available at http://www.math.ku.dk/sjo/papers/ReducedRankRegression.pdf.

[8] Doornik, J. A. and R. J. O. Brien. 2002. 'Numerically Stable Cointegration Analysis'. *Computational Statistics and Data Analysis* 41 (1): 185–193.

# Part II
# Cloud Computing Applications for Smart Grid

**CHAPTER 5**

# Demand Response

Typically, in traditional power delivery systems, energy is supplied to the customers using non-renewable energy sources. Therefore, the energy generators are based on fossil fuels. Consequently, carbon emission in the environment is very high, and there is potentially a great impact on the environment. In contrast, in the smart grid system, energy is supplied to the end-users using renewable and non-renewable energy sources. Additionally, the implementation of distributed energy sources plays an important role for real-time energy management in the smart grid. Consequently, in the presence of distributed energy sources at the customers' end, electricity consumers can schedule their *shiftable* appliances to the *off-peak* hours to relieve the extra load on the grid during *on-peak* hours. This mechanism is widely known as demand response (DR). The objective of any demand response mechanism is to reduce energy consumption cost to the customers and reduce imbalance between the energy supply and demand, while modifying the energy consumption pattern according to real-time price signals.

## 5.1 Fundamentals of Demand Response and Challenges

Different technologies are introduced in a smart grid system in order to establish the demand response mechanism – demand scheduling, online energy monitoring, and deployment of distributed energy sources. As mentioned earlier, customers can schedule their shiftable appliances to off-peak hours so as to minimize their energy consumption cost. Additionally, the real-time energy demand to the grid can also be relieved during peak hours. In the traditional power delivery system, such strategies cannot be deployed by the service providers in order to schedule the customers' energy demand. Therefore, we have on-peak and off-peak hours in which energy demand is more and less, respectively, than the energy supply to the grid. On the other hand, using the demand response mechanism, the energy demand from the customers can be modified according to the energy supply to the grid in order to minimize the imbalance between the energy supply and the demand to the grid. However, there are several challenges that should be considered in order to install demand response mechanisms in real-life environments.

1. *Peak load reduction:* In the smart grid, there are three different time-slots depending on the energy load on the grid – *on-peak*, *mid-peak*, and *off-peak*. During on-peak hours, energy demand from the customers is high. On the other hand, during mid-peak and off-peak hours, demands from the customers are moderate and low, respectively. The smart meters, which are deployed at the customers' end, send the real-time demand information to the grid. In such a scenario, for a particular area, large numbers of smart meters are expected to be deployed to get real-time information about the customers' demands. The grid needs to process this huge information before deciding the actual properties of the current time-slot, i.e., whether it is on-peak, off-peak, or mid-peak. After deciding the state of the current time-slot (i.e., on-peak, off-peak, and mid-peak), the grid needs to take adequate decisions in order to minimize the peak load from the grid. For example, adequate pricing policy, intergration of self-generated energy sources, and other energy sources can be integrated to balance the real-time energy supply and demand to the grid. However, such decision taking mechanism is not an easy task to execute.

2. *Integration of renewable energy sources:* Self-generated energy sources play an important role in minimizing the peak load in smart grids. However, renewable energy sources (such as solar and wind energy) are intermittent in nature. Therefore, availability of such energy sources is quite uncertain in real time. Therefore, heavily relying on renewable energy sources may affect the reliability of energy-service to the end-users.

3. *Heterogeneity of the customers:* In smart grids, different entities such as residential customers, third-party users, and vehicle users take part in real-time energy management, and thus,it is required to use a common platform of energy-service. In such a setting, it is difficult to support such heterogeneous entities in a single platform. For example, the demand response strategy implemented for residential customers may not support mobile users (such as plug-in electric vehicle users).

## 5.2 Different Demand Response Mechanisms

In this section, the different mechanisms that are needed for establishing demand response in smart grids are discussed. We categorize the existing technologies according to their types – *economic*, *emergency*, and *ancillary* demand response. The economic demand response is essentially the energy cost minimization of the customers and the service providers. On the other hand, emergency demand response is the measure of reliability of the energy-service to the customers during load-shedding. Emergency generators are used to avoid load-shedding conditions in order to maintain the reliability of the energy-service. The reliability of energy-service is measured according to the ratio of energy supplied and energy demanded by customers (maximum value of energy reliability is 100%). Finally, ancillary demand response introduces security measures of energy-service in the smart grid. In this section, we discuss these different demand response mechanisms.

## 5.2.1 Economic demand response

The economic demand response can be implemented at both ends – customers' and service providers'. At the customers' end, the economic demand response takes into account changes in electricity usage according to real-time prices in order to minimize electricity consumption cost. On the other hand, at the service providers' end, economic demand response is the maintenance of real-time energy supply and demand to provide reliable energy-services so as to relieve the extra load from the grid during peak-hours. Different economic demand response schemes (without cloud) are described in the following subsections.

### Multiple utility providers and customers

In the smart grid, typically, two types of entities exist – customers and service/utility providers (grid). The objective of the customers is to minimize the energy consumption cost, while fulfilling energy requirements. On the other hand, the objective of the service providers is to maximize their revenues by providing electricity to the customers, while considering customers' participation in energy trading. Therefore, a demand response scheme is adequate if it supports such requirements for the customers and the service providers as well. In addition to this, in the smart grid architecture, multiple customers expect to be served by multiple service providers. In such a scenario, interactions between customers and service providers are introduced in order to achieve demand response in the smart grid [1]. The service providers set their own prices of unit energy to maximize their payoffs. On the other hand, according to the prices set by the service providers, a customer chooses the optimal (best) service provider for which his/her energy consumption cost is minimized. Consequently, each customer has multiple options to choose the appropriate service provider, which, in turn, implies that a service provider decides real-time price considering other service providers' strategies in order to take into account customers' participation. Further, in order to take appropriate decisions on both sides, use of game theoretic demand response mechanisms can be applied [1]. In the game theoretic solution approach, bi-directional communication facility is utilized for successful interaction between the customers and the grid. For example, let us assume that three homes have a monthly electricity consumption of 150 kWh, 250 kWh, and

125 kWh. Tariffs for electricity consumption are set by different service providers as follows – (a) $7.5 per kWh for up to 100 kWh and $10 per kWh for above 100 kWh; (b) $8 per kWh for upto 150 kWh and $12 per kWh for above 150 kWh; and (c) $10.5 per kWh for up to 250 kWh, and $14 per kWh for above 250 kWh. Therefore, the customer with energy demand 150 kWh should consume energy from service provider two and the customer with energy demand 250 kWh should consume energy from service provider three, and so on, in order to minimize the electricity consumption cost.

**Demand scheduling**
Demand scheduling is a promising approach for customers to minimize energy consumption cost and to establish an economic demand response. On receiving information about the real-time price from the service provider, the customer compares his/her *satisfactory* price with the real-time price. On getting a satisfactory price from the service provider, the customer consumes electricity to fulfill his/her energy requirements. On the other hand, if the price is greater than the satisfactory price, the customer schedules his/her appliances (shiftable appliances only) to other time-slots to minimize the energy consumption cost. Therefore, demand scheduling plays an important role in the realization of demand response in the smart grid. Additionally, another possible approach can be time-slot-based energy consumption [2]. As discussed earlier in Section 5.1, there are three different time-slots in the smart grid – on-peak, mid-peak, and off-peak. During on-peak hours, the real-time price is high, as the demand from the customers to the grid is high. On the other hand, real-time price is moderate and low in the mid-peak and on-peak hours, respectively. Therefore, the customer consumes energy in the current time-slot if the corresponding time-slot is off-peak. On the other hand, the customer schedules appliances (shiftable) to off-peak hours if the corresponding time-slot is on-peak with a certain delay incurred by the customer that can be up to a possible extent. Additionally, the customer consumes energy for non-shiftable appliances to fulfill requirements. Such schemes are centralized (i.e., price in different time-slots are pre-defined by the service provider, irrespective of real-time supply and demand to the grid), and scheduling of appliances are done by day-ahead pricing policy. On the other hand, opportunistic scheduling can be incorporated for an economic demand response mechanism [3]. In such a scenario, appliances are automatically scheduled according to the real-time price. The scheme determines the best time to use a particular appliance, and therefore, schedules that appliance in the specific time-slot.

**Integration of plug-in hybrid electric vehicles**
Integration of plug-in hybrid electric vehicles (PHEVs) is one of the important features of the smart grid. The PHEVs charge their batteries during off-peak hours, and act as energy storage devices. They discharge their batteries to the grid during on-peak hours to relieve the extra load on the grid; during these hours, they act as energy source devices. Therefore, optimal charging and discharging schemes have great impact on the feasibility

of incorporating PHEVs in the smart grid. Several PHEV charging and discharging schemes are also proposed in the literature for the smart grid [1], [4], [5]. Micro-grids decide real-time price based on either the energy supply and demand or only the energy demand to them. On receiving the real-time price from multiple micro-grids, PHEVs choose the optimal one for which their charging costs are minimized. Additionally, PHEVs also discharge their batteries to the optimal micro-grid for which their revenue is maximized. In such a scenario, optimal demand response policy can be established with the incorporation of PHEVs in the smart grid.

## 5.2.2 Emergency demand response

The distributed architecture of the smart grid divides the electric distribution side into smaller cells, which are called micro-grids. Micro-grids provide electricity to the customers from a combination of renewable and non-renewable energy sources. Emergency demand response needs to be implemented due to load-shedding, higher energy demand than supply, and load variations. Some of the existing emergency demand response schemes are discussed in this section.

**Integration of storage devices**

Integration of storage devices in the smart grid is a promising approach to support emergency demand response in the smart grid. The storage devices are used as energy buffers at the micro-grid side, and participate in real-time energy management. During on-peak hours (i.e., energy demand is higher than the available supply to the micro-grid), energy from available storage devices relieves the extra demand on the micro-grid by providing stored electricity, and, thus, these devices balance the real-time supply-demand curve for providing reliable energy-service to the customers. Therefore, storage devices help to provide reliable energy-service to customers. Due to the intermittent behavior of self-generated energy sources (such as solar and wind), energy demand and supply is highly irregular (i.e., supply from self-generated energy sources at the micro-grid and demand for self-generated energy sources at the customers' end fluctuates). Customers can use their storage energy during on-peak hours to minimize the energy consumption cost. Thus, storage devices at the customers' and micro-grids' ends support emergency demand response.

**Integration of emergency generators**

In order to handle exigent situations (such as blackouts and absence of renewable energy) in the smart grid, each micro-grid is expected to have emergency generators installed at the distribution side. The emergency generators are used occasionally to balance the real-time supply–demand curve, when there is a scarcity of supply to the micro-grids or blackouts [6]. Micro-turbines are mostly used as emergency generators in the smart grid; they act as small-scale distributed generators.

### 5.2.3 Ancillary demand response

In contrast to the *economic* and *emergency* demand response, ancillary demand response is concerned with reliability and scalability of energy-services. The simplest way to minimize the peak load is demand scheduling. However, sometimes a large number of generators may fail and there is a huge demand in a small time-scale; smart grids needs an adequate demand response policy that can take care of such odd situations [7]. In the presence of ancillary demand response, micro-grids can handle such situation and provide reliable energy supply to the customers.

## 5.3 Problems with Existing Approaches without Cloud

With the growing demand of the smart energy-service, the number of smart meters is growing enormously day-by-day to meet the requirements of the customers. The service provider needs to access and process the real-time data generated from the smart meters in order to take optimal decisions. However, this is a significantly challenging task to the service provider with the dedicated systems due to the following reasons.

**Various functionalities of micro-grids**
The micro-grids provide electricity to the customers in a distributed manner. Therefore, they decide the real-time price depending on the supply or demand (or both), so as to establish efficient energy management. However, a micro-grid needs to have sufficient information of other micro-grids in order to take appropriate decisions (such as energy exchange among micro-grids and real-time market price), which is dependent on the entire smart grid environment. The distributed architecture, that is, the existing infrastructure without a cloud environment, may not support such requirements. Additionally, micro-grids are also expected to work in the island mode when there is a breach of security service. In order to meet this requirement, it is crucial to deploy an early-warning system in the smart grid.

**Variation in customers' requirements**
As previously discussed, there are three different time periods in the smart grid – on-peak, mid-peak, and off-peak. Therefore, the resource required to support the customers' need is variable. Consequently, it may be cost-expensive to have a dedicated system for the smart grid as the dedicated system must support on-peak as well as off-peak requirements.

**Appliance scheduling**
The major concern with appliance scheduling schemes is the control over appliances. It is not clear who will control the appliances – customers or service providers. If the customers schedule their appliances according to their requirements, then the corresponding time-slots may be converted to on-peak hours from off-peak hours, which,

in turn, renders the policy cost-expensive rather than cost-effective. On the other hand, if the service providers schedule the appliances, then there are two major concerns. First, privacy of the customers may not be maintained. Second, the changes in the appliances' running times must be informed to the customers immediately. Therefore, an efficient mechanism needs to be deployed in order to take into account such issues.

### Issue with single failure in the distribution side
In the smart grid, energy is distributed through a hierarchical process – master–slave architecture [5]. A single failure in the master–slave architecture may lead to a complete blackout situation in the smart grid due to the communication gap between two levels in the hierarchical process.

### Needs to support multiple customers
In the smart grid, different customers – residential, third-party, and PHEVs – are expected to play important roles for real-time energy management. Therefore, real-time information must be available online to the customers for better monitoring of their home appliances. The customers must be able to choose an optimal service provider, though micro-grids provide electricity in a distributed manner. However, the existing infrastructure (without cloud) experiences many difficulties in supporting multiple customers in a cost-effective manner [9].

## 5.4 Cloud-Based Demand Response in Smart Grid

In this section, we discuss different demand response schemes for smart grid energy management with cloud computing applications. Additionally, we also discuss different demand response schemes that can be applied for energy management in the data center used to fulfill smart grid requirements. Therefore, cloud-based demand response technologies are discussed in two different contexts – residential energy management and data center energy management. Figure 5.1 presents a schematic view of the cloud-based demand response model for the smart grid. The customers send their energy consumption information to the price response from the cloud infrastructure. The generation unit sends real-time supply information to the cloud-based systems. The service provider requests the cloud server for real-time demand response on the basis of real-time energy consumption information. According to the real-time supply and demand information, the cloud responds to the service provider with a suitable demand response mechanism. Therefore, with the presence of cloud-based systems, the supply and demand sides maintain-time energy production and consumption, respectively, in order to establish the demand response model for the smart grid.

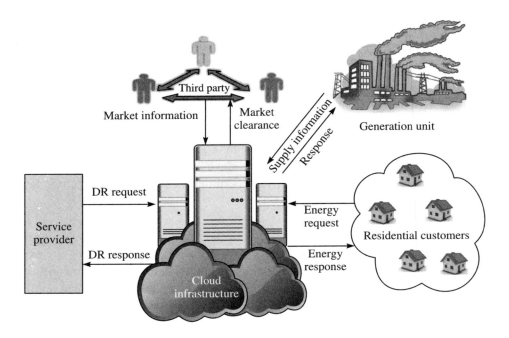

**Figure 5.1** Cloud-based demand response model for the smart grid

## 5.4.1 Demand response in smart grid energy management

**Cloud-based PHEVs management**

Real-time energy demand frequency regulation is an important aspect of the smart grid in order to provide electricity to the end-users in a reliable manner. Due to the high carbon footprint of fossil fuels, environment-friendly energy sources are gaining popularity in the context of smart grid energy generation and distribution. Consequently, distributed storage devices and plug-in hybrid electric vehicles participate in real-time energy management. However, it is required to have real-time information to (or from) the PHEVs for reliable energy exchange between the PHEVs and the grid. Additionally, it is important to consider data center energy consumption, while optimizing PHEVs' energy management based on real-time information. Therefore, cloud-based PHEVs management schemes can be considered for reliable and cost-effective operation in the smart grid. Using cloud-based schemes, we have both information – data center energy consumption cost for interaction with PHEVs and the PHEVs' energy consumption cost. Consequently, a joint optimization scheme is useful to consider both energy and frequency regulation in the smart grid. A cloud-based optimization problem for PHEVs management is presented in [10]. The optimization is done in a hierarchical process. First, a controller organizes the energy among the cloud servers, storage units, and PHEVs. Then, the second layer called the 'capacity planning layer' controls the capacity in response to the market price. Figure 5.2 presents the schematic view of the two-layered proposed scheme for cloud-based PHEVs management (adopted from [10]).

In every fixed time interval, the data center receives a power budget from the power allocator. According to the power budget, the data center follows the real-time CPU regulation so as to minimize frequency regulation. The minimization of the changes of power regulation at the data centers can be represented as an optimization problem as follows [10]:

$$\underset{f_i(k), i \leq N_a}{\text{Minimize}} |\hat{P}_s(k) - P_s(k)|$$

subject to

$$f_i(k) \in F_i \tag{5.1}$$

$$R(k) \leq R_s \tag{5.2}$$

where $P_s(k)$ is the real-time power consumption, $\hat{P}_s(k)$ is the power budget, $f_i(k)$ is the aggregated frequency, which must be less than or equal to the maximum allowable frequency, and $R(k)$ is the response time from the data center, which must be less than or equal to the threshold response time threshold response time, $R_s$. $N_a$ is the total number of entities taking part in demand response, such as servers $(N_s)$, storage units $(N_u)$, and PHEVs $(N_p)$, i.e., $N_a = N_s + N_u + N_p$. Similarly, the PHEVs also receive a power budget from the power controller and charge their batteries in real-time. Therefore, the objective

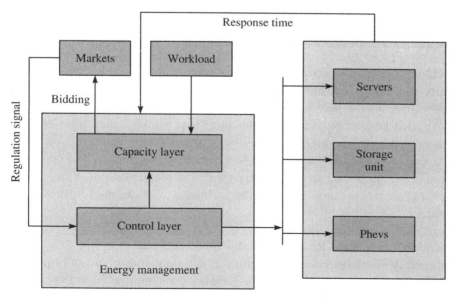

**Figure 5.2** Two-layer hierarchical cloud-based demand response

of the PHEVs is also minimization of the difference between the power budget and the real-time power consumption. Consequently, with the presence of the control layer, data centers and PHEVs minimize the real-time energy consumption cost. Additionally, the capacity layer controls the charging process of the PHEVs batteries within a deadline period.

**Real-time online energy management**
Self-generated energy sources are becoming popular in order to establish a green smart grid environment. However, due to the intermittent behavior of the renewable energy sources, it is difficult to predict real-time energy supply from such energy sources. Therefore, micro-grids need to schedule their energy generators – renewable and non-renewable – in order to provide uninterrupted energy supply to the customers. Consequently, in such a scenario, an online algorithm is useful to schedule the energy generators in a micro-grid's area [11]. The objective of the micro-grids is to minimize energy generation cost with the availability of different generators. Mathematically, the objective function can be defined as follows:

Minimize $C(y,w,o,h)$
subject to

$$0 \leq o(t) \leq o_{\max}(t) \tag{5.3}$$

$$0 \leq h(t) \leq h_{\max}(t) \tag{5.4}$$

$$y_i \in [0,1] \tag{5.5}$$

$$0 \leq w(t) \leq w_{\max}(t) \tag{5.6}$$

The objective function $C(y,w,o,h)$ is a cost function which should be minimized in order to minimize the energy consumption cost. $y_i$ denotes that the generator $i$ is either ON or OFF, $h(t)$ denotes the heat power that can be less than or equal to a maximum value $h_{\max}(t)$, $o(t)$ denotes the output at the generator side which is less than or equal to the maximum value $o_{\max}(t)$, and $w(t)$ denotes the power obtained from the main grid which is also less than or equal to the maximum power can be obtained $w_{\max}$. The online algorithm is evaluated in a cloud-based server, which provides an efficient solution for scheduling the generators. Figure 5.3 presents a schematic architecture of the micro-grid with different energy sources.

**Communication mechanisms for cloud-based demand response**
In cloud-based demand response, the communication network plays an important role in optimizing the energy usage and minimizing energy consumption cost. Typically, we have a two-tier architecture for cloud-based demand response – edge cloud and core cloud. In the edge cloud, the cloud takes decisions based on local knowledge. For example, a

micro-grid takes decisions based on the energy consumption information from its own customers. On the other hand, in the core cloud, information from all micro-grids (edge clouds) is aggregated, and a coordinated decision is taken in a centralized manner. Figure 5.4 depicts a two-stage cloud-based demand response model. In such a model, it is important to study the communication mechanisms for cloud-based demand response. As discussed in Chapter 1, the information from smart meters is collected at data aggregator units (DAUs), and the aggregated information is further forwarded to the service providers through multi-hop or single-hop communication paths. In the multi-hop scenario, a DAU sends its aggregated data to another DAU and so on, and eventually, the data reach to the service provider. On the other hand, in the single-hop scenario, a DAU directly sends the aggregated information to the service provider. In the communication network, if $\eta_k$ is the bit error rate of the communication channel in the region of the DAU $k$ and it follows Bernoulli function, then the packet error probability $p_k(s_m)$ with packet size $s_m$ is represented as follows:

$$p_k(s_m) = 1 - (1 - \eta_k)^{s_m} \tag{5.7}$$

where $s_m$ is the packet size. On successful reception of the message, the receiver sends an acknowledgement (ACK) to the sender. However, if there is any error in the transmission, the packet is re-transmitted to improve the packet delivery ratio. In the latter case, the

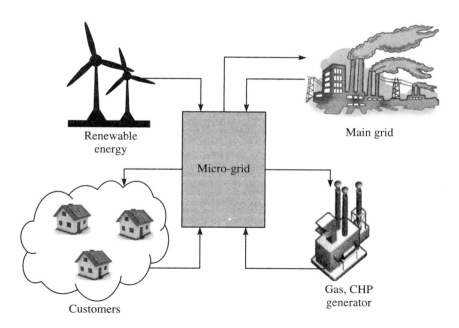

**Figure 5.3** Micro-grid providing electricity to customers with different energy generators

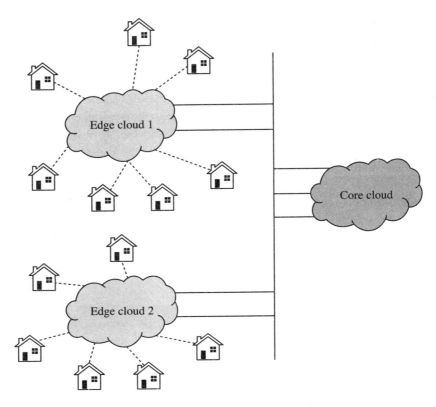

**Figure 5.4** Two-stage cloud-based demand response model

receiver sends an automatic repeat request (ARQ) error message to the sender. Consequently, the average number of re-transmissions required for a bit error rate $\eta_k$ with packet size $s_m$ is as follows:

$$p_{\text{re-trans}} = \frac{p_k(s_m)}{1 - p_k(s_m)} \tag{5.8}$$

Therefore, reliable communication network is required in order to have efficient smart grid energy management.

To send a message from a customer to the edge cloud, three different approaches can be followed – (a) hop by hop; (b) end-to-end; and (c) intermediate cache; the approaches are discussed as follows [12].

- *Hop-by-hop*: In the hop-by-hop model, the information of a sender is forwarded to the receiver through intermediate nodes. Therefore, on detecting unsuccessful transmission, all the intermediate nodes re-transmit. This approach minimizes the delay in packet delivery. However, it increases the cost, as all the intermediate nodes store, buffer, and send the messages repeatedly. Mathematically, the number of transmissions through intermediate nodes can be represented as:

$$N_{HbH}(i,H,s_m) = \sum_{k=i}^{i+H-1} \frac{1}{1-p_k(s_m)} = \sum_{k=i}^{i+H-1} (1-\eta_k)^{-s_m} \quad (5.9)$$

where $N_{HbH}(i,H,s_m)$ denote the number of transmissions from a source node $i$ to a destination node with packet size $s_m$ through $H$ intermediate nodes.

- *End-to-end*: In the end-to-end approach, the sender node stores, buffers, and sends unacknowledged packets. Therefore, it reduces the cost; however, the packet delivery delay increases. It is noteworthy that energy consumption cost for communication increases as the distance is very high in the end-to-end approach. Similar to the hop-by-hop transmission, the number of transmission in the end-to-end approach can be represented as:

$$N_{EtE}(i,H,s_m) = \frac{H}{1 - \sum_{j=i}^{i+H} p_j(s_m) \prod_{k=i}^{j-1}(1-p_k(s_m))} \quad (5.10)$$

- *Intermediate cache*: This is a combination of hop-by-hop and end-to-end approaches, i.e., a subset of intermediate nodes store, buffer, and send the unacknowledged packets, instead of all intermediate nodes and the sender node doing the same. Therefore, it maintains a trade-off between packet delivery delay and the associated cost. Some of the intermediate nodes are selected as cache points that re-transmit the lost packets requested by the receiver. At a region $r$ with $C_r$ cache points, the following equation is satisfied:

$$H_r = (C_r - 1)\Delta_r + 1 \quad (5.11)$$

where $\Delta_r$ is the hop distance from two nodes. Similar to the hop-by-hop and end-to-end approaches, the number of transmissions is calculated as follows:

1. when $i < H_r$:

$$N_{IC}(i,H,s_m) = N_{EtE}(i,\Delta_r - A, s_m) + N_{HbH}(H_r, 1, s_m)$$

$$+ \sum_{k=B+2}^{C_r-1} N_{EtE}((k-1)\Delta_r + 1, \Delta_r, s_m) \quad (5.12)$$

where $A$ and $B$ are defined as: $A = \mod(i-1, \Delta_r)$ and $B = \lfloor (\frac{i-1}{\Delta_r}) \rfloor$.

2. When $i = H_r$:

$$N_{IC}(i, H, s_m) = N_{HbH}(H_r, 1, s_m) \quad (5.13)$$

Finally, the total bandwidth required to transfer $M$ packets is represented as follows:

$$B = \sum_{r=1}^{R} \sum_{j=1}^{D_r} N_{rj} \sum_{m=1}^{M} s_m N_{\text{method}}(j, H_r - j + 1, s_m) \quad (5.14)$$

where $N_{rj}$ and $D_r$ denote the number of users and aggregators in the region $r$. $N_{\text{method}}$ represents the delivery method of the message, which can be either hop-by-hop or end-to-end or intermediate cache based.

### 5.4.2 Demand response in data centers for the smart grid

Due to the on-growing interests in cyber physical systems (CPS), the energy consumed by the data centers also increases day-by-day. As a result, carbon emission to the environment increases, which, in turn, poses a challenge to researchers for maintaining a green environment. Similarly, smart meters are expected to be deployed at the distribution side for all customers in order to communicate with service providers in the smart grid. Therefore, the information generated from the smart meters needs to be stored and processed at the data center. Consequently, the data center consumes large amount of energy to process such information. Therefore, the idea is to use demand response schemes in the data center for cost-effective cyber physical systems for the smart grid, where cyber refers to the data centers and physical refers to the power grid. Some of the existing demand response models for smart grid data center energy management are presented in this section.

**Dynamic power management in distributed data centers**

To support smart grid requirements, Internet data centers (IDCs) play an important role in information processing and management [13]. Therefore, energy consumed by the IDCs is very high, so as to provide online reliable services to the smart grid. An energy-efficient demand response scheme was proposed by Wang et al. [13] for minimizing the energy consumption cost for the data centers. In such a model, the proposed scheme dynamically manages the power consumed by the data centers. The data centers respond to the market price decided by the service provider, and according to the price, the IDCs change their energy consumption pattern dynamically. Consequently, the data center operators balance their energy consumption, while fulfilling the requirements from the users (such as customers, service providers, and third parties). One of the possible approaches can be taking optimal decisions based on the dynamic work-load and power supply from the smart grid to the IDCs. Therefore, the optimization problem of the corresponding scenario can be formulated as follows [13]:

$$\underset{w_i,\rho_{ji}}{\text{Minimize}}\ T_{\text{total}} = \sum_{i=1}^{N} w_i P_i(w_i) E_i$$

subject to

$$\sum_{i \in N} w_i \leq W_i \tag{5.15}$$

$$\sum_{i=1}^{N} \rho_{ji} \leq R_j \tag{5.16}$$

where $w_i$ is the number of functional servers for the $i^{\text{th}}$ location in the smart grid, which is less than or equal to the all the servers available for the $i^{\text{th}}$ location. $P_i(w_i)$ is the price incurred by the functional servers, and $E_i$ is the energy consumption for one server. Additionally, Equation (5.15) denotes that the number of functional servers must be equal to or less than the maximum number of servers; $\rho_{ji}$ is the assigned rate of functioning of each server, and it is always less than or equal to the maximum rate $R_j$. Therefore, according to the real-time price and energy consumption, the system dynamically assigns workload to the individual servers to minimize the energy consumption cost. The aforementioned optimization problem can be converted to a quadratic programming model, and thus, it can be solved accordingly. In such a scenario, each data center (IDC) uses real-time energy according to their requirements, which, in turn, changes the real-time price. Therefore, the IDC acts as a real-time price decider in the electricity market.

**Demand response management in data centers**
A coordinated demand response mechanism is also proposed [14]. In such a scenario, the objective of the scheme is to minimize energy consumption by the Internet data centers (IDCs), so as to minimize the carbon footprint in the environment. All the IDCs are homogeneous in nature in terms of the operating frequency, while maintaining reliability and quality-of-service. Depending on the real-time situation, the IDCs schedule their task in order to improve the environmental impacts.

Figure 5.5 illustrates that the data centers are deployed as distributed systems, and respond to the customers' requirements locally. The data centers can also communicate and dispatch their individual loads to other data centers, depending on their available resources. The service providers also interact with the data centers for real-time demand response policy.

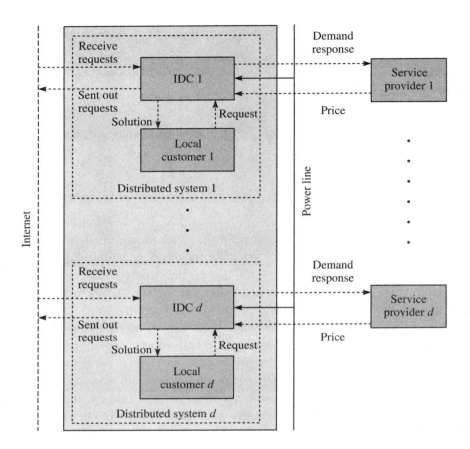

**Figure 5.5** Electric demand response in data centers

**Cost minimization with multi-source power supply (SmartDPSS)**

Due to the growing demand from the users, increasing energy consumption cost and carbon emissions pose a challenge to the cloud service providers in order to provide cost-effective operation. In order to mitigate such problems, data centers are equipped in such a way that they can be operated with different energy sources – renewable, non-renewable, and storage. Consequently, efficient use of such different energy sources is also a big challenge to the cloud service provider. An efficient operating scheme, called SmartDPSS, for use by the data centers with the availability of different energy sources was proposed by Deng et al. [15]. SmartDPSS controls the use of the different energy sources online without requiring advance knowledge of the availability of the energy sources.

Table 5.1 shows a brief comparison among different demand response approaches with their limitations.

Table 5.1  Comparison of different demand response approaches

| Approaches | Methodology | Limitations |
|---|---|---|
| PHEVs Management | Two-layer architecture for PHEVs management in smart grid. In the first layer, a controller controls the energy distribution among cloud server, storage units, and PHEVs. In the second layer, the total energy capacity is controlled while considering market price. | Appliances are expected to be controlled from the service providers. Therefore, security and privacy of the customers are important issues that need to be considered. |
| Online Energy Management | Online scheduling of renewable and non-renewable energy generators in a micro-grid's area. Therefore, the minimization of energy generation cost is studied depending on the availability of different energy generators. | Intermittent behavior of renewable energy sources needs to be considered. Prediction-based algorithm can be used in such cases. |
| Data Center Energy Management | Minimization of energy consumption by cloud data centers so as to minimize the carbon footprint. | Homogeneous behavior of the data centers is assumed. In a practical scenario, data centers may be heterogeneous in nature. |

## 5.5 Future Trends and Issues

The cloud-based demand response model minimizes energy consumption cost incurred by the customers. Additionally, such a model helps the service providers to provide energy-services to the customers in a cost-effective and reliable manner. Cloud computing methods can be used for optimal demand response toward cost-efficient and reliable energy-services to the end-users. However, there are still several issues that need to be addressed for realizing such models before large-scale deployment. Some of major challenges are as follows.

### Security and privacy

Due to information-centric energy management, privacy of the users' information is a matter of serious concern. In the smart grid, different users (such as customers, third-parties, and service providers) take part in real-time energy management. Therefore, customers' personal information (such as energy consumption pattern) may be accessible to third parties. Consequently, it is required to include adequate security policies, so that users' privacy is preserved. This is a really challenging task for service providers while uploading information on to the cloud platform.

### Control of appliances

As discussed earlier, service providers can also control customers' appliances depending on real-time requirements such as reduction in peak load. However, it is not clear yet who will be acting as the primary controller of the appliances, and who as the secondary controller. With the invention of smart grid technologies, customers expect uninterrupted energy supply. However, due to such appliance control mechanisms, customers may face interrupted power supply.

## 5.6 Summary

In this chapter, a detailed discussion of the state-of-the-art smart grid demand response mechanism is presented. First, we discussed different demand response models – *economic*, *emergency*, and *ancillary*. Thereafter, different existing demand response models without cloud applications were summarized. Then, problems with these existing ones were also discussed. To overcome the problems in demand response mechanisms without cloud applications, different demand response mechanisms based on cloud computing applications were illustrated. It is evident that cloud-based mechanisms are useful to establish cost-effective and reliable demand response model in the smart grid architecture. Finally, future directions of the existing cloud-based demand response models were presented with the several issues that need to be addressed.

### Test Your Understanding

Q01. What is demand response?
Q02. What are on-peak and off-peak hours?
Q03. What are the three demand response mechanisms?
Q04. What is the objective of demand response?
Q05. What is SmartDPSS?

Q06. What are the challenges that should be considered while installing an available scheme for demand response in the practical environment?

Q07. Briefly explain the different demand response mechanisms.

Q08. What are the problems with the existing demand response approaches without cloud?

Q09. Briefly describe cloud-based PHEVs management.

# References

[1] Maharajan, S., Q. Zhu, Y. Zhang, S. Gjessing, and T. Basar. 2013. 'Dependable Demand Response Management in the Smart Grid: A Stackelberg Game Approach'. *IEEE Transactions on Smart Grid* 4 (1): 120–132.

[2] Erol-Kantarci, M. and H. T. Mouftah. 2011. 'Wireless Sensor Networks for Cost-efficient Residential Energy Management in the Smart Grid'. *IEEE Transactions on Smart Grid* 2 (2): 314–325.

[3] Yi, P., X. Dong, A. Iwayemi, C. Zhou, and S. Li. 2013. 'Real-time Opportunistic Scheduling for Residential Demand Response'. *IEEE Transactions on Smart Grid* 4 (1): 227–234.

[4] Misra, S., S. Bera, and T. Ojha. 2015. 'D2P: Distributed Dynamic Pricing Policy in Smart Grid for PHEVs Management'. *IEEE Transactions on Parallel and Distributed Systems* 26 (3): 702–712.

[5] Fan, Z. 2012. 'A Distributed Demand Response Algorithm and Its Application to PHEV Charging in Smart Grids'. *IEEE Transactions on Smart Grid* 3 (3): 1280–1290.

[6] Saha, A. K., S. Chowdhuri, S. P. Chowdhuri, and P. A. Crossley. 2008. 'Micro-turbine Based Distributed Generator in Smart Grid Application'. In *IET Seminar Report on Smart Grids for Distribution*. Frankfurt, Germany: IET.

[7] Ma, O., N. Alkadi, P. Cappers, P. Denholm, S. Dudley, J. Goli, M. Hummon, S. Kiliccote, J. MacDonald, N. Matson, D. Olsen, C. Rose, M. Sohn, M. Starke, B. Kirby, and M. O'Malley. 1995. 'Demand Response for Ancillary Services'. *IEEE Transactions on Smart Grid* 4 (4): 1988.

[8] Bera, S., S. Misra, and J. J. Rodrigues. 2015. 'Cloud Computing Applications for Smart Grid: A Survey'. *IEEE Transactions on Parallel and Distributed Systems* 26 (5): 1477–1494.

[9] Kim, H., Y. J. Kim, K. Yang, and M. Thottan. 2011. 'Cloud-based Demand Response for Smart Grid: Architecture and Distributed Algorithms'. In *Proc. of the IEEE SmartGridComm*, Brussels. pp. 398–403.

[10] Li, S., M. Brocanelli, W. Zhang, and X. Wang. 2014. 'Integrated Power Management of Data Centers and Electric Vehicles for Energy and Regulation Market Participation'. *IEEE Transactions on Smart Grid*. 5 (5); 2283–2294.

[11] Lu, L., J. Tu, C. K. Chau, M. Chen, and X. Lin. 2013. 'Online Energy Generation Scheduling for Micro-grids with Intermittent Energy Sources and Co-generation'. In *Proc. of the ACM SIGMETRICS* pp. 53–66.

[12] Yaghmaee, M. H., A. L. Garcia, M. Moghaddassian. 2017. 'On the Performance of Distributed and Cloud-Based Demand Response in Smart Grid'. *IEEE Transactions on Smart Grid*. doi: 10.1109/TSG.2017.2688486.

[13] Wang, P., L. Rao, X. Liu, and Y. Qi. 2011. 'Dynamic Power Management of Distributed Internet Data Centers in Smart Grid Environment'. In *Proc. of the IEEE GLOBECOM*. pp. 1–5.

[14] Chen, Z., L.Wu, and Z. Li. 2014. 'Electric Demand Response Management for Distributed Large-Scale Internet Data Centers'. *IEEE Transactions on Smart Grid* 5 (2): 651–661.

[15] Deng, W., F. Liu, H. Jin, and C. Wu. 2013. 'SmartDPSS: Cost-Minimizing Multi-Source Power Supply for Datacenters with Arbitrary Demand'. In *Proc. of the IEEE 33rd Intl. Conf. on Distributed Computing Systems*. pp. 420–429.

# CHAPTER 6

# Geographical Load-Balancing

The distributed architecture of smart grid facilitates micro-grids to provide electricity to the end-users in a distributed manner. Additionally, the micro-grids provide electricity to the customers using a combination of renewable and non-renewable energy sources. Therefore, each micro-grid has distributed generation facilities (such as combined heat power, solar, and wind), which act as the renewable and non-renewable energy sources. Consequently, micro-grids maintain a balance between real-time energy supply and demand to them. In order to have real-time information of the supply and demand, data centers need to be deployed for each micro-grid with bi-directional communication facility. In such a scenario, distributed load-balancing among the energy sources and data servers is required. Load-balancing is a technique by which massive load on a particular micro-grid can be reduced significantly.

## 6.1 Need for Load-Balancing in Smart Grid

There is a need to implement load-balancing in a smart grid due to the following reasons.

- As micro-grids provide electricity to customers in a distributed manner, the energy management policy implementation also takes place in a distributed manner at the micro-grids' end. Therefore, it may happen that in a particular micro-grid's area, demand from the customers is very high. Consequently, real-time energy load on that micro-grid is very high, and it has to provide adequate electricity to the customers in order to meet the energy requirements of the latter. As a result, the micro-grid has to buy electricity from the main grid (through a market pricing policy) in order to meet the requirements. However, at the same time, it may happen that real-time demand is very low in another micro-grid. Therefore, there is an energy excess in that micro-grid. Consequently, it sells the surplus energy to the main grid. In such an energy exchange mechanism, a massive amount of energy is lost due to transmission loss. However, with the implementation of the load-balancing scheme, service areas of the micro-grids can be dynamically changed in order to maintain the balance between real-time energy supply and demand.

- Similar to the energy scarcity problem in a micro-grid's area, real-time information propagation may lead to the creation of a massive data load on the servers. With the help of bi-directional communication in the smart grid, customers communicate with the service providers for energy consumption. Therefore, real-time information propagation at the data centers also increases due to the massive energy demand from the customers. However, with dedicated servers for different micro-grids, it is quite difficult to meet such requirements of data processing. Therefore, it is useful to have a load-balancing scheme at the data centers in order to meet the fluctuating demands from the customers.

## 6.2 Challenges

There are different challenges that need to be addressed before one can implement any load-balancing scheme in a smart grid. The different challenges are as follows.

- *Billing policy*: Smart meters are deployed at the customers' end in order to calculate real-time energy consumption cost according to the real-time energy price and energy demand from the customers. Therefore, it is important to deploy adequate billing mechanisms, while allowing the customers to consume energy from different micro-grids, as price of energy from different micro-grids may be different [1]. Otherwise, it will be a challenge for the customers to schedule their demands in different time-slots.

- *Reliability to the customers*: It is also important to consider the reliability of the energy-service to the customers, while allowing load-scheduling among the micro-grids. The ultimate goal of the smart grid is to provide electricity to the customers in a reliable and cost-effective manner. Therefore, load-balancing schemes should not cause interruption in the energy-service provided to the customers. For example, a micro-grid decides to exchange energy with another micro-grid, depending on real-time supply and demand situations. However, the sudden high demand for energy on the micro-grid may lead to interrupted energy supply. Therefore, such uncommon situations also need to be addressed while implementing load-balancing schemes in the smart grid.

- *Security and privacy*: With the implementation of load-balancing, information processing can take place at any of the available servers. On the other hand, different service providers may have different pricing and billing policies. Therefore, security and privacy of the customers must be maintained while allowing multiple service providers to work on a common platform.

## 6.3 Problems with Existing Load-Balancing Approaches without Cloud

### 6.3.1 Coalition formation

As mentioned earlier, micro-grids distribute electricity in a distributed manner, and can also exchange energy with other micro-grids (or the main grid). In contrast to the energy exchanging process among different micro-grids, it is beneficial to change the service area of the micro-grids according to the real-time energy supply and demand to them. Dynamically controlling the service areas of the micro-grids using coalition formation[1] is one of the cost-effective schemes in the smart grid. In this context, one possible approach is game-theoretic service area selection in order to reduce loss in the micro-grids [13]. Similarly, cooperative energy distribution network can also be formed using coalitional game theory [6].

The micro-grids form coalitions among themselves based on real-time information of supply and demand from the customers. However, this formation is a daunting task without adequate information from both the sides–the grid and customers. Therefore, communication architecture and data servers are important components of the smart grid [7]. There is a need to have a cost-effective load-balancing scheme for the smart grid, in which adequate information should be available from anywhere and anytime. Only the implementation of distributed servers at each micro-grid may not be sufficient to handle such issues.

### 6.3.2 Flexible demand forecasting

In contrast to coalition formation, another possible approach is that customers forecast their demands in different time-slots in advance to the grid. According to the forecasted demand from the customers, the service providers deploy suitable energy management schemes in the smart grid – this is known as the economic dispatch problem. Additionally, real-time demand from customers can be rescheduled to other time-slots if the real-time demand is more than the supply. Therefore, customers follow the micro-grids' responses in order to deploy the load-balancing mechanism in the smart grid. Accordingly, they defer their demand from the selected time-slot to other time-slots [10].

---
[1] Coalition formation is conceptualized as multiple users form a group with common interests. Therefore, we can consider different aspects and instrumental parameters to form coalitions among multiple users.

In such a scenario, all the demands from the customers are assumed to be flexible, and they can be rescheduled at any time. However, in a realistic scenario, all demands are not flexible, i.e., they cannot be rescheduled randomly to different time-slots. Moreover, with the traditional load forecasting schemes, there is a major chance of errors. Load forecasting errors may lead to unreliable energy-services to customers; the revenue of the service providers may decrease as well. Lower demand forecasting leads to higher energy scarcity in real-time and higher demand forecasting increases the system cost [3]. Therefore, in order to have a realistic scenario and an adequate load forecasting mechanism, cloud-based approaches may be the most cost-effective and reliable method.

### 6.3.3 Centralized load controller

Load controllers are introduced to balance real-time energy supply and demand. The load controller is installed at the service providers' end. It acts as a centralized device in the smart grid for real-time energy management. In such a scenario, the load controller only controls the shiftable appliances (such as washing machine) in every intra time-slot [11]. Depending on the real-time demand from customers and available supply, the controller controls the modes (ON or OFF) of shiftable appliances in order to maintain the supply–demand curve. The controller turns OFF appliances during high demand from customers and turns ON appliances during low demand from the customers so as to maintain cost-effective energy supply to the customers.

The deployment of the centralized controller means that the distributed architecture of the smart grid may not be supported. Additionally, due to centralized control policies, customers may not get certain services, though they wish to have them. Consequently, a distributed control policy is required based on real-time supply and demand information for load-balancing in the smart grid.

## 6.4 Cloud-Based Load-Balancing

We see that there are several problems in introducing load-balancing in smart grids using traditional approaches (without cloud). Consequently, in order to overcome such problems, cloud-based load-balancing schemes are introduced in the smart grid. The inherent feature, infrastructure as a service (IaaS), of cloud computing technology provides the flexibility to distributed energy and data load-balancing in the smart grid. In the smart grid, different distributed energy sources and consumers take part in real-time energy trading. Additionally, such energy sources and consumers are distributed geographically in different regions, and have different impacts on real-time energy management. Data centers are also geographically distributed over different locations. In order to provide reliable energy-services to the customers, the service controller dynamically routes the requests to an optimal data center. In addition to this, data centers may also replicate similar data to provide required services from different locations, which may increase the cost and wastage of storage memory. Moreover, a bi-directional communication architecture of the smart grid gathers real-time information generated from smart meters and service providers. This real-time information must be stored and processed in order to take adequate decisions. Consequently, data load-balancing at the Internet data centers (IDCs) is one of the important aspects that need to be considered for reliable energy management in the smart grid. Cloud-based load-balancing schemes have two-fold advantages – energy demand balancing, and real-time information storage and processing – that minimizes the energy consumption cost and provides reliable energy to the customers in the smart grid. Different cloud-based load-balancing schemes are discussed in this section.

### 6.4.1 Price-based energy load-balancing

Dynamic pricing policy according to the availability of real-time energy supply and demand from the customers allows the smart grid to balance real-time energy consumption by the customers [1]. On the other hand, distributed data centers also consume a major part of the energy in order to support the smart grid requirements [12], [14]. Therefore, the real-time energy demand is increased in those micro-grids' areas where data centers are deployed. Consequently, real-time price in such micro-grids increases, and, thus, the operating cost of such data centers, as incurred by the service providers, is high. In order to overcome such a cost-expensive information processing scheme, data centers can be operated in a distributed manner, where the real-time price is low [14]. As discussed in Chapter 5 (demand response), micro-grids decide real-time price depending on the real-time energy demand or supply or both supply and demand. Therefore, during a particular time period, the real-time price is different for all the micro-grids. In such a scenario, the service provider assigns real-time task to the data center, for which the real-time energy consumption cost is low. In addition to this, the price of energy varies as a factor of 10 from one time period to another [9]. Therefore, after a certain duration of time, if the real-time energy consumption cost for a particular data center is high, the service provider pools the task from that server and assigns a new one for which the energy consumption cost is low. Figure 6.1 illustrates such a dynamic service scheduling scheme in the smart grid in order to deploy load-balancing in smart grid data centers. In such a load-balancing scheme, two controllers are introduced – a smart grid controller and a cloud computing controller. The smart grid controller sets the real-time price of energy with an aim to maximize revenue from energy distribution. On the other hand, the cloud controller controls the distributed location-based price signals to the data centers in order to maximize its revenue. For example, data from data center 1 is pooled by the service provider, and is assigned to data center 3, while considering the energy prices at both the locations – data centers 1 and 3.

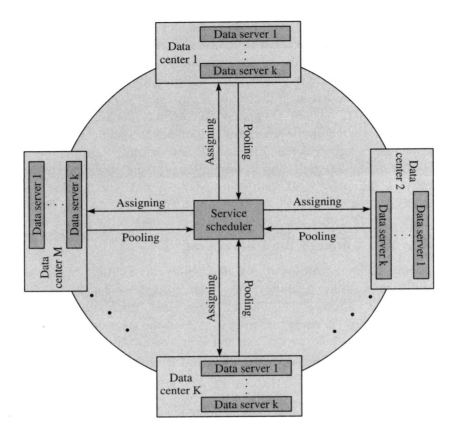

**Figure 6.1** Service pooling and assigning for load-balancing in data centers

### 6.4.2 Load-balancing at the smart grid data centers

The load-balancing scheme discussed in Section 6.4.1 considers only the queuing delay while transferring a task from one data center to another. However, transmission delay is another important factor that needs to be considered while implementing load-balancing in smart grid data centers [14, 15, 16]. Figure 6.2 presents an overview of a data center powered by smart grid (adopted from [16]). Different data centers are allocated to serve particular requests from the customers by coupling the transmission delay and permitted delay for a particular request. Figure 6.3 shows the schematic diagram of the system. First, the front-end server checks the permitted delay and the corresponding transmission delay for a particular data center. Then, the front-end server allocates the requests to the appropriate data center, while maintaining the service level agreements (SLAs). Before presenting the optimization problem of such IDC-based systems to support smart grid infrastructure, it is required to calculate the delay factors (such as transmission and queuing delay), which are crucial for load-balancing.

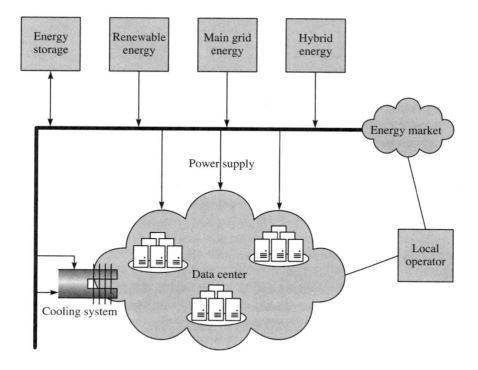

**Figure 6.2** Schematic view of a data center powered by smart grid

- *Transmission delay*: According to the scheme (presented in Figure 6.3), the front-end servers allocate a particular request from a customer to an optimal IDC. Therefore, the allocation of tasks to the IDCs results in transmission delay. In real systems, we have the routing delay for request distribution as the transmission delay. Mathematically, the transmission delay can be presented as follows:

$$T_d(t) = f(\lambda_{ij}(t)) \tag{6.1}$$

where $\lambda_{ij}(t)$ is the request distribution factor and $f(\cdot)$ is the function used to derive the request distribution parameter, $\lambda_{ij}$, for calculating the transmission delay from the $i^{\text{th}}$ server to the $j^{\text{th}}$ data center.

- *Queuing delay*: The queuing delay at the data center can be expressed as follows:

$$Q_d(t) = \frac{1}{m_i(t)\eta_i(t) - \sum_{i=1}^{F}\lambda_{ij}(t)}, \quad \lambda_{ij}(t) \in N \tag{6.2}$$

where $\eta_i(t)$ is the request rate from the front-end servers to the data center $i$ at time $t$, and $F$ is the number of front-end servers.

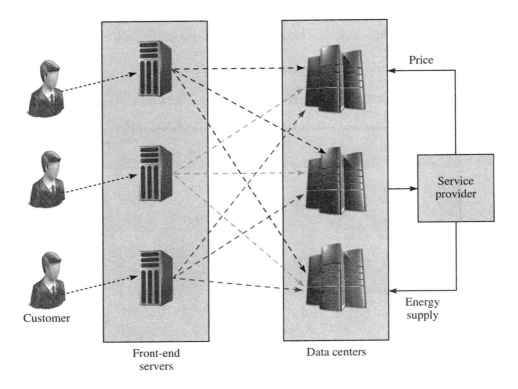

**Figure 6.3** Schematic diagram of a transmission delay-based load-balancing using front-end servers

The objective of the service provider is to minimize the energy consumption cost for the data centers, while fulfilling the customers' requirements [15]. Therefore, the minimization of the energy consumption cost to the service provider can be presented as an optimization problem, as follows.

$$\text{Minimize} \sum_{i=1}^{N} w_i(t) E_i(t) p_i(t)$$

subject to

$$0 \leq w_i(t) \leq S_i \tag{6.3}$$
$$0 \leq E_i(t) \leq E_i^{\max} \tag{6.4}$$

where $w_i(t)$ is the number of working servers at the data center $i$, $E_i(t)$ is the amount of energy consumed by each server, and $p_i(t)$ is the real-time price decided by the electric utility. Additionally, the number of working servers, $w_i(t)$, is always less than the number of available servers, which is positive. The energy consumed by each server, $E_i(t)$, also has a maximum value and is positive. The amount of energy consumed by a server can be calculated according to its operating frequency, as follows:

$$E_i(t) = a_i f_i^{co} + b_i \tag{6.5}$$

where $a_i$ and $b_i$ are the predefined constants, $f_i$ is the operating frequency, and $co$ is the coefficient parameter, which varies in the range 2.5 to 3 in real-life scenarios [15]. Therefore, the optimization problem can be redefined as follows:

$$\text{Minimize} \sum_{i=1}^{N} w_i(t)(a_i f_i^{co} + b_i) p_i(t)$$

subject to

$$T_d(t) + Q_d(t) \leq D_d(t) \tag{6.6}$$

$$0 \leq w_i(t) \leq S_i \tag{6.7}$$

$$0 \leq E_i(t) \leq E_i^{max} \tag{6.8}$$

$D_d(t)$ is the permissible delay by the customers to meet their service level agreements (SLAs). The optimization problem can be solved using a heuristic-based branch and bound method [15].

### 6.4.3 Renewable energy-aware load-balancing

Renewable energy sources have resulted in the creation of green smart grid architecture. In Chapter 5 (demand response), we saw that renewable energy sources are used to support cost-effective energy supply to the customers. Additionally, powering the data centers with renewable energy sources can reduce the carbon footprint in the environment significantly. Therefore, incorporating renewable energy sources may be cost-effective for distributed load-balancing [17]. In contrast to non-renewable energy sources (such as fossil fuel), renewable energy sources (such as solar and wind) are location dependent. Additionally, energy produced by renewable energy sources is highly intermittent. Consequently, the electric utility provider decides the real-time energy price depending on the available renewable energy sources, which is different for different geographical regions. In such a renewable energy-based pricing scenario, the service provider schedules their data centers accordingly in order to minimize the energy consumption cost. Mathematically, the minimization of energy consumption cost of the data center operation in the presence of renewable energy sources can be presented as follows:

$$\text{Minimize} \sum_{t \in T} \sum_{i \in N} p_{nr}(t) m_i(t) + \sum_{t \in T} \sum_{i \in N} p_r(t) r_i(t)$$

subject to

$$0 \leq m_i(t) \leq N \tag{6.9}$$

$$0 \leq r_i(t) \leq N \tag{6.10}$$

$$m_i(t) + r_i(t) = N_{\text{req}}(t) \tag{6.11}$$

where $p_{nr}(t)$ and $p_r(t)$ are the real-time price for non-renewable and renewable energy sources, $m_i(t)$ and $r_i(t)$ are the number of non-renewable and renewable energy sources, $N_{\text{req}}(t)$ is the number of required energy sources.

### 6.4.4 Load-balancing at data center networks

Similar to the residential and other customers, data centers also consume energy to process information coming from multiple sources. These data centers can be deployed and managed in a distributed manner like the micro-grids. Therefore, geographical energy load-balancing at the data center networks is required to make the latter energy-efficient and sustainable. From this perspective, a distributed stochastic geographical load-balancing approach is introduced at the cloud data center networks [20]. Figure 6.4 presents a schematic diagram of such a load-balancing model.

In the model, users send their requests to an intermediate node, known as the *mapping node* (MP). On receiving the requests, an MP forwards the requests to a suitable data center depending on the user's requirement, communication and networking cost, load at individual data center, and energy usage. Therefore, MPs act as gateway devices between users and data centers. These MPs and data centers are geographically located at different places. For example, multiple data centers may be deployed at different places in a country. In such an approach, two types of workload are considered – delay-tolerant workload (DW) and interactive workload (IW). DW is similar to shiftable load, which can be queued for some time. On the other hand, IW needs to be served as soon as possible. The objective of the system is to minimize operational cost while considering associated revenue generated from the energy distribution. The operational cost takes into account the following: energy transmission cost, conventional generator cost, cost for discharging battery, and cost for energy distribution. On the other hand, the revenue captures the following scenarios: (a) revenue for serving DW and (b) revenue for serving the IW. Several constraints, such as battery capacity, total generation and workload, and queue length of DW are also considered. For simplicity, they are not discussed in detail as the main objective is to provide an insight into the proposed model. Interested readers may refer to [20]. In the optimization problem, there are few constraints having an infinite time horizon, which leads to shaping up a multi-dimensional optimization problem. The dimension reduction method is useful to reduce the dimension of the optimization problem in order to get a solution in polynomial time.

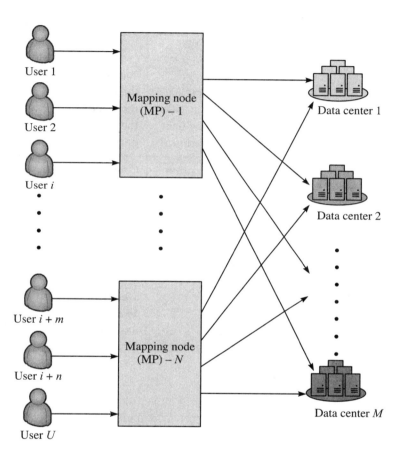

**Figure 6.4** Schematic view of geographical load-balancing at data center network

Table 6.1 presents a brief comparison of different load-balancing approaches in the smart grid with their limitations.

**Table 6.1** *Comparison of different load-balancing approaches*

| Approaches | Methodology | Limitations |
|---|---|---|
| Price-based load-balancing | Real-time data processing is done in the data center for which the energy consumption cost to the service provider is low. This requires dynamic task allocation for real-time data processing in the smart grid. | Communication cost and throughput are the important factors that need to be considered. Otherwise, the service provider may incur more cost, though the real-time price in the particular area is low. |
| Data center load-balancing | Dynamic service allocation is done according to the transmission delay and maximum allowable delay. | Only transmission delay is considered for service allocation. |
| Renewable energy-aware load-balancing | Powering the data centers with renewable energy sources. | Renewable energy sources are dependent on weather conditions. Therefore, weather condition-based load-balancing can be studied. |

## 6.5 Future Trends and Issues

- Price-based load-balancing for the smart grid data center is widely implemented world-wide. However, other parameters – bandwidth cost, performance – are also needed to be jointly studied in the smart grid context with energy consumption cost [9, 18]. By forming the joint optimization problem (with energy parameters and network parameters), it is possible to have a complete data operation scheme in the smart grid in a cost-effective and reliable manner.
- Similar to price-based load-balancing, weather-based load-balancing can also be done in the smart grid [9]. It is observed that data centers are distributed over different geographic locations. Therefore, at different locations, weather conditions are also different. Consequently, the environment has great impact on the energy consumption cost minimizing parameters such as cooling.

- In addition to the energy consumption cost minimization, it is necessary to look into the environmental impacts for different cost-effective solutions. The growing demand for cyber-physical systems (CPS) increases carbon footprint in the environment, which is a serious issue that needs to be addressed by researchers. Overall, it is necessary to have a green environment with cost-effective energy consumption.
- The cost related to task assignment (pooling) to (from) a data center from (to) another data center has major impact on the overall energy consumption cost. Frequent transfers may consume massive energy, which, in turn, makes the scheme cost-expensive rather than cost-effective. Therefore, task scheduling mechanisms should consider both the real-time energy price and cost involved in the transfer process.
- It is also important to consider intermittent behavior of the renewable energy sources, while integrating them into the smart grid data center operations. Otherwise, a small interruption in the data center operation may lead to overall system failure.

## 6.6 Summary

In this chapter, different load-balancing schemes required for smart grid system were discussed. First, different challenges that need to be considered before one can implement a load-balancing scheme were presented. Thereafter, existing load-balancing schemes (without cloud operation) and their shortcomings were discussed in order to support the cost-effective distributed architecture of the smart grid. Different existing cloud-based load-balancing schemes were also discussed. Additionally, different future opportunities of cloud-based schemes and related issues were also presented in the context of the smart grid. Finally, we discussed how cloud-based schemes are useful to establish load-balancing schemes in order to meet smart grid requirements.

### Test Your Understanding

Q01. What do you mean by load-balancing?
Q02. Why do we need load-balancing in smart grid?
Q03. What are the challenges that need to be kept in mind before implementing any load-balancing scheme in smart grid?
Q04. Briefly describe coalition formation.
Q05. How can flexible demand forecasting be a problem with existing load-balancing approach without cloud?

Q06. What is a centralized load controller?

Q07. What are the advantages of cloud-based load-balancing?

Q08. What is meant by price-based energy load-balancing?

Q09. What are the two delay factors that are important for implementing a load-balancing scheme?

Q10. What are the advantages of renewable energy aware load-balancing?

Q11. What are the limitations of renewable energy aware load-balancing?

Q12. What is the limitation of load-balancing in data center?

Q13. What are the limitations of price-based load-balancing?

Q14. What is a CPS?

Q15. What is the disadvantage of CPS in the context of load-balancing and energy consumption?

Q16. What is the objective of an energy-service provider in smart grid?

Q17. What is the difference between transmission delay and queuing delay?

Q18. Point out some future trends and issues of load-balancing.

# References

[1] Misra, S., S. Bera, and T. Ojha. 2015. 'D2P: Distributed Dynamic Pricing Policy in Smart Grid for PHEVs Management'. *IEEE Transactions on Parallel and Distributed Systems* 26 (3): 702–712.

[2] Wang, Y., X. Lin, and M. Pedram. 2013. 'A Sequential Game Perspective and Optimization of the Smart Grid with Distributed Data Centers'. In *Proc. of the IEEE PES ISGT*. pp. 1–6.

[3] Tomic, S. D. 2013. 'A Study of the Impact of Load Forecasting Errors on Trading and Balancing in a Micro-grid'. In *Proc. of the IEEE Green Technologies Conference*. pp. 443–450.

[4] Rahman, A., X. Liu, and F. Kong. 2014. 'A Survey on Geographic Load Balancing Based Data Center Power Management in the Smart Grid Environment'. *IEEE Communications Surveys and Tutorials* 16 (First Quarter): 214–233.

[5] Bera, S., S. Misra, and J. J. Rodrigues. 2015. 'Cloud Computing Applications for Smart Grid: A Survey'. *IEEE Transactions on Parallel and Distributed Systems* 26 (5): 1477–1494.

[6] Saad, W., Z. Han, and V. H. Poor. 2011. 'Coalitional Game Theory for Cooperative Micro-Grid Distribution Networks'. In *Proc. of the IEEE ICC*. pp. 1–5.

[7] Sernec, R., M. Rizvic, E. Usaj, M. Sterk, S. Strozak, P. Nemcek, and I. Sauer. 2014. 'Communication Architecture for Energy Balancing Market Support on Smart Grid'. In *Proc. of the IEEE ENERGYCON*. pp. 1500–1508.

[8] Wang, Y., X. Lin, and M. Pedram. 2014. 'Coordination of the Smart Grid and Distributed Data Centers: A Nested Game-based Optimization Framework'. In *Proc. of the IEEE PES ISGT*. pp. 1–5.

[9] Qureshi, A., R. Weber, H. Balakrishnan, J. Guttag, and B. Maggs. 2009. 'Cutting the Electricity Bill for Internetscale Systems'. *ACM SIGCOMM Computer Communication Review* 39 (4): 123–134.

[10] Hassan, N. U., X. Wang, S. Huang, and C. Yuen. 2013. 'Demand Shaping to Achieve Steady Electricity Consumption with Load Balancing in a Smart Grid'. In *Proc. of the IEEE PES ISGT*. pp. 1–6.

[11] Zhang, Y. and N. Lu. 2013. 'Demand-side Management of Air Conditioning Cooling Loads or Intra-hour Load Balancing'. *IEEE Transactions on Smart Grid* 4 (4): 2100–2108.

[12] Guodarzi, H. and M. Pedram. 2013. 'Geographical Load Balancing for Online Service Applications in Distributed Datacenters'. In *Proc. of the IEEE Cloud Computing (CLOUD)*. pp. 351–358.

[13] Wei, C., Z. Fadlullah, N. Kato, and A. Takeuchi. 2014. 'GT-CFS: A Game Theoretic Coalition Formulation Strategy for Reducing Power Loss in Micro Grids'. *IEEE Transactions on Parallel and Distributed Systems* 25 (9): 2307–2319.

[14] Rao, L., X. Liu, L. Xie, and W. Liu. 2010. 'Minimizing Electricity Cost: Optimization of Distributed Internet Data Centers in a Multielectricity-Market Environment'. In *Proc. of IEEE INFOCOM*. pp. 1–9.

[15] Shao, H., L. Rao, Z. Wang, Z. Wang, X. Liu, and K. Ren. 2014. 'Optimal Load Balancing and Energy Cost Management for Internet Data Centers in Deregulated Electricity Markets'. *IEEE Transactions on Parallel and Distributed Systems* 25 (16): 2659–2669.

[16] Chen, T., Y. Zhang, X. Wang, and G. B. Giannakis. 2016. 'Robust Workload and Energy Management for Sustainable Data Centers'. *IEEE Journal on Selected Areas in Communications* 34 (3): 651–664.

[17] Paul, D. and W. D. Zhong. 2013. 'Price and Renewable Aware Geographical Load Balancing Technique for Data Centers'. In *Proc. of the IEEE ICICS*. pp. 1–5.

[18] Zheng, X. and Y. Cai. 2011. 'Reducing Electricity and Network Cost for Online Service Providers in Geographically Located Internet Data Centers'. In *Proc. of IEEE/ACM International Conference on Green Computing and Communications (GreenCom)*. pp. 166–169.

[19] Li, J., Z. Li, K. Ren, and X. Liu. 2012. 'Towards Optimal Electric Demand Management for Internet Data Centers'. *IEEE Transactions on Smart Grid* 3 (1): 183–192.

[20] Chen, T., A. G. Marques, and G. B. Giannakis. 2017. 'DGLB: Distributed Stochastic Geographical Load Balancing over Cloud Networks'. *IEEE Transactions on Parallel and Distributed Systems* 28 (7): 1866–1880.

# CHAPTER 7

# Dynamic Pricing

With the implementation of bi-directional communication facility, service providers have real-time information about energy consumption by the customers and energy production by the generators. In such cases, dynamic pricing based mechanisms are more appealing than traditional static pricing based ones. Service providers decide real-time price based on the demand from customers or supply from the generation side or both the supply and demand. Consequently, customers have the opportunity to use electricity according to their preferred real-time price decided by the grid. Additionally, service providers maintain the real-time supply–demand curve by controlling the real-time price. When demand increases, the price may be increased to reduce the extra load from the grid, and vice-versa. Customers also schedule their energy consumption depending on the real-time price decided by the grid. Therefore, dynamic pricing policy is one of the important features of the smart grid that provides dynamic load balancing in the smart grid.

## 7.1 Deployment of Dynamic Pricing in Smart Grids

As the name suggests, in dynamic pricing, the price of energy varies over time according to the energy supply or demand or both. To implement the dynamic energy pricing policy, we need to address several challenges that will be discussed in this section.

### 7.1.1 Determination of actual time-slot

We discussed earlier that there are three different time-slots in the smart grid – on-peak, mid-peak, and off-peak. Based on the past days' behavior, state (on, mid, or off-peak) of a particular time-slot is determined. For example, the office hour can be considered as on-peak, the morning and evening can be considered as mid-peak, and night can be considered as off-peak. However, in real-time, the situation in the grid may be different. For example, energy demand on a particular day may be different from another day. Moreover, as the smart grid is equipped with renewable energy sources, which are intermittent in nature, energy supply to the grid may also vary over multiple days. Therefore, it is not possible to blindly say that a particular time period is mid-peak, or on-peak, or off-peak. Consequently, it is important to define suitable strategies to determine the state of a time-slot.

## 7.1.2 Need for adequate infrastructure

Dynamic pricing policy depends on real-time information from customers and generators. Smart meters play an important role in providing real-time information to service providers. The demand information of the customers is relayed with the help of data aggregator units (DAUs) to the base station. However, due to information loss in the communication networks, service providers may not have adequate information. For smart grid, there is, as of now, a lack of clear road-map of the communication technologies to be used. Particularly, whether the existing communication technology would be used or new technologies would be employed is not clear. If the existing technology is to be used, it is necessary to consider the fact that millions of smart meters will take part in the energy trading process. Consequently, the existing communication technology should be capable of handling such massive data generated from the smart meters. In contrast, if new technology would be used, specifying the architecture for that is a big question. Otherwise, we may have partial information at the service providers' end, which may lead to inefficient pricing policy in the smart grid. Therefore, adequate communication network needs to be deployed in order to implement dynamic pricing policy in the smart grid.

## 7.2 Existing Dynamic Pricing Policies without Cloud

In smart grid, there are different pricing policies discussed in literature that calculate real-time energy consumption cost increased by the customers and revenue to the service providers. We discuss different existing pricing policies without cloud computing applications with their shortcomings in the subsequent sections.

## 7.2.1 Day-ahead pricing policy

Day-ahead pricing policy in the smart grid is static in nature. The entire day is divided into different time-slots (e.g., each slot of 1 hour duration) [13]. Depending on the customers' energy consumption pattern of the past days, the service provider defines the energy price for different time-slots as on-peak, off-peak, and mid-peak. A time-slot is treated as an on-peak hour if the price in that time-slot is higher than a pre-defined value. On the other hand, in the off-peak and mid-peak hours, the price is below a pre-defined lower threshold value and is between the higher and lower threshold values, respectively. Customers in the smart grid consume energy according to the pre-defined real-time price, and also schedule their appliances in different time-slots. However, in the smart grid, real-time supply and demand are expected to be dynamic in nature. In day-ahead pricing scheme, the dynamic behavior of the smart grid is not considered. Consequently, with the pre-defined constant pricing policy, real-time demand to the grid may be increased due to lower price in a particular time-slot. Hence, the day-ahead static pricing policy may lead to an imbalance between energy supply and demand.

### 7.2.2 Demand-based pricing policy

Another possible strategy is the demand-based real-time pricing policy, in which price of energy depends on the energy requested by the customers. In such a pricing scenario, real-time price can be determined based on the demand from individual customers, i.e., usage-based dynamic pricing (UDP) [12]. Mathematically, the demand-based real-time energy price ($p_t$) is represented as follows:

$$p_t = \begin{cases} p_1 & \text{if } D_{c,t} \leq D_{th,1} \\ p_2 & \text{if } D_{c,t} > D_{th,1} \text{ and } D_{i,t} \leq D_{th,2} \\ a + bD_{i,t} + cD_{i,t}^2 & \text{if } D_{c,t} > D_{th,1} \text{ and } D_{i,t} > D_{th,2} \end{cases} \quad (7.1)$$

where $p_1$ and $p_2$ are pre-defined static prices and $D_{c,t}$ denotes the total demand from a community. On the other hand, $D_{th,1}$ and $D_{th,2}$ denote two pre-defined thresholds for energy demand, and $D_{i,t}$ denotes the energy demand from the $i^{th}$ customer. Finally, $a$, $b$, and $c$ are pre-defined constants. In such a pricing policy, the energy price is proportional to the energy demand from the customer. Consequently, the cost incurred by the customer is high with higher demand, and vice-versa. However, it may happen that the demand from a customer is high, as a consequence of which the real-time energy price is high, though the total demand to the grid is very less. This, in turn, leads to inefficient energy management in the smart grid. Similarly, a customer pays for the reduced amount of energy consumption cost with lower demand, though total demand to the grid is very high, which also leads to inefficient energy management in the smart grid. Therefore, implementation of policies purely based on individual customers may not be adequate to determine the real-time pricing policy.

## 7.2.3 Supply-based pricing policy

Similar to the demand-based pricing policy, real-time price can also be evaluated based on the supply to the grid. In such a scenario, the real-time price is evaluated by adopting a quadratic cost function (QCF) based on the total supply to the grid [10, 6]. Mathematically, supply-based real-time price $(p_t)$ is represented as follows:

$$p_t = aS_t^2 + bS_t + c \qquad (7.2)$$

where $S_t$ is the total energy supply to the grid, and $a$, $b$ and $c$ are pre-defined constants. Similar to the case of demand-based pricing, dynamic behavior of the smart grid should also be considered. It may happen that the total supply is high, which may result in high real-time price. This, in turn, causes less participation of the customers in energy consumption. On the other hand, real-time price may be less due to the less amount of energy production, which, in turn, increases the peak-load on the grid due to high demand from the customers.

## 7.2.4 Supply–demand-based pricing policy

Lastly, another possible strategy to determine real-time price is to consider both the supply and demand. In such a scenario, the grid decides the real-time price of energy based on the real-time supply to the grid and demand from the customers [1]. In a particular time, if the total supply is higher than the total demand from the customers, the real-time price is lower than the base-price, in order to attract customers to consume more energy, thereby minimizes energy loss. On the other hand, real-time price is higher than the base-price when the total demand from the customers is higher than the total supply to minimize customers' demand, and thereby relieve the demand from the grid for reliable energy management. Mathematically, it is represented as follows:

$$p^t = p_b + \{\tan^{-1}(e^\lambda) - \gamma\} \qquad (7.3)$$

where $\lambda$ is the difference between the energy demand, $\mathcal{D}_t$, and energy supply $\mathcal{S}_t$, and can be written as $\lambda = (\mathcal{D}_h^t + \mathcal{D}_r^t) - \mathcal{S}_t$, and $\gamma$ is a pre-determined constant. $\mathcal{D}_h^t$ and $\mathcal{D}_r^t$ denote the *home* and *roaming* energy demand from customers, respectively. The home energy demand is conceptualized as the energy demand to a micro-grid from users who are registered to the former. Therefore, home users are static in nature; the micro-grid can get a day-ahead energy consumption pattern for these users. In contrast, the roaming energy demand is considered as the energy demand to a micro-grid from users who are not registered with the former, such as the traditional mobile communication system. Consequently, the micro-grid experiences sudden increase in the energy demand to it in the presence of energy demand from roaming users. With an increase in the supply, $\mathcal{S}_t$, the real-time price, $p^t$, decreases, while the demand, $\mathcal{D}_t$, from the customers is either fixed or decreases. On the other hand, the real-time price increases with an increase in the demand, $\mathcal{D}_t$, from the customers, whereas the supply, $\mathcal{S}_t$, is either fixed or decreases. Consequently, customers are interested in consuming more energy when the price is low, and vice-versa. Such a dynamic pricing policy illustrates a well-balanced pricing scheme that maintains the supply–demand curve while considering customers' participation. However, in such a pricing scenario, the grid always requires adequate information of both supply and demand. Without cloud-based model, it may be difficult to get such information to the grid adequately.

## 7.3 Problems with Existing Approaches without Cloud

The aforementioned pricing policies can be used to determine real-time price of energy, depending on the energy supply or demand, or both the supply and demand. However, there are a few limitations with the existing distributed pricing policies as will be discussed in this section.

### 7.3.1 Local knowledge of supply–demand information

As the energy demand and supply are calculated in a distributed manner at the micro-grid level, the grid always has local information of real-time energy supply and demand. Consequently, the real-time price of energy differs between micro-grids. Thus, the local knowledge of energy supply–demand information may lead to local optima, which may not be effective in establishing a well-balanced smart grid environment.

### 7.3.2 Unfair pricing tariffs for customers

In addition to only local knowledge about the energy supply–demand information, the customers in a smart grid environment may suffer from unfair pricing tariffs. The micro-grids decide the real-time price of energy in a distributed manner, depending on their individual status. Therefore, micro-grids may set different values for the pre-defined constants according to their service level agreements (SLAs). Consequently, this may lead to unfair pricing tariffs to the customers in a smart grid environment.

To overcome such limitation, cloud-based pricing policies are proposed in the literature [3, 5, 13].

## 7.4 Cloud-Based Dynamic Pricing Policies

In this section, different cloud-based pricing policies are discussed, which are useful to adderss the above mentioned limitations (i.e., the problem without cloud).

Internet data centers (IDCs) are large-volume consumers in the electricity market [3]. They have a great impact on the real-time energy market policies. Consequently, the presence of IDCs in a smart grid environment should be considered while deciding real-time energy price in addition to the regular energy market. Towards this objective, Wang et al. [13, 5] proposed an energy cost minimization problem in a smart grid energy market, while considering the impact of IDCs on it. Consequently, interactions between cyber systems (data centers) and physical systems (power grid) were studied by the authors. Similar to other existing demand response mechanisms, IDCs can also change their energy demand in response to the real-time energy market, in order to minimize the energy consumption cost. In such a model, market clearing price is denoted as follows:

$$P(Q_{idc}) = \begin{cases} \alpha_1 Q_{idc} + \beta_1, & Q_{idc} \leq Q_0 - Q_c \\ \alpha_2 Q_{idc} + \beta_2, & Q_{idc} > Q_0 - Q_c \end{cases} \quad (7.4)$$

where $P(Q_{idc})$ denotes the market clearing price while considering the energy demand from IDCs. $Q_{idc}$, $Q_0$, and $Q_c$ denote the energy demand from IDCs, a pre-defined *knot* demand, and energy demand from other consumers in the smart grid, respectively. $\alpha_1$, $\beta_1$, $\alpha_2$, and $\beta_2$ are pre-defined constants. Finally, the real-time energy market clearing price problem can be formulated as a quadratic cost function that can be solved using the non-linear programming approach. The main advantage is that, in such a scheme, the service providers have complete energy demand information from both the conventional consumers and the IDCs. Therefore, we have a globally optimized energy market price, which can be treated as fair, depending on the energy demand.

In a smart grid environment, multiple consumers are expected to communicate with the service providers. Moreover, they are also expected to communicate among themselves in order to consume electricity in a cooperative manner. Therefore, huge amount of energy is also expected to be consumed in the communication process. Consequently, the energy consumption cost for communication is also an important factor to consider while deciding the real-time price of energy. Li et al. [9] proposed such a dynamic pricing strategy considering the energy consumption for communication and computation of tasks, in addition to the conventional electricity consumption. Mathematically,

$$P = P_{comp} + P_{comm} \quad (7.5)$$

where $P_{comp}$ and $P_{comm}$ denote the total energy price for computation and communication to execute a task. Further, $P_{comp}$ and $P_{comm}$ are represented as follows:

$$P_{comp} = \sum_{i=1}^{N} E_{pr,i} l_i P_t \quad (7.6)$$

and

$$P_{\text{comm}} = \sum_{i=1}^{N}\sum_{j=1}^{N}\sum_{k=1}^{R}\sum_{m=1}^{R} E_{i,j}^{\text{comm}} I_{i,k} I_{j,m} r_{k,m} P_{t,E_{i,j}^{\text{comm}}} \qquad (7.7)$$

where $E_{pr,i}$, $l_i$, $P_t$ denote the energy required for processing the $i^{\text{th}}$ task, the length of the task, and the real-time price at time $t$, respectively. On the other hand, $E_{i,j}^{\text{comm}}$ and $r_{k,m}$ denote the energy consumption for communication, and the communication ratio between two processors $k$ and $m$, respectively. Finally, $I_{i,k}$ and $I_{j,m}$ denote the binary indicator variables. Therefore, the consumer can schedule their tasks considering both the computation and communication costs, in order to minimize the overall energy consumption cost. Energy management units (EMUs) schedule the tasks in an adaptive manner, while leveraging the global information of the smart grid environment.

Similar to the demand-based electricity pricing scheme, Narayan et al. [8] proposed a power-aware pricing policy for cloud service provisioning. The real-time price of cloud services is decided based upon the amount of service requests from the users. The price is high if the demand is high, while keeping the capacity of the cloud fixed, and vice-versa. Mathematically, they have modeled a pricing policy for cloud service provisioning as follows:

$$P_i^t = \phi + k\left(u_i \times \frac{C_t}{C} - \alpha\left(C - \sum_i u_i\right)\right) \qquad (7.8)$$

where $P_i^t$ is the price for cloud instance $i$ at time $t$, $\phi$ is a fixed cost for utilizing the cloud instance, $k$ and $\alpha$ are predefined constants, $C_t$ denotes the total operational cost, $C$ denotes the resource capacity of the system, and $u_i$ is the resource utilization by the cloud instance $i$. In such a pricing scheme, cloud service provisioning captures the following [8]:

- The cost for license and maintenance of the cloud platform is not incurred by the service providers. All such costs can be taken into consideration by the fixed price ($\phi$), and it is split equally among all the cloud instances.
- The price of using the cloud instance is low when the correlation factor $(C - \sum_i u_i)$ is high, i.e., demand is lower than the cloud capacity.
- On the contrary, when the correlation factor is low, price is high. Therefore, a good balance between demand and supply of cloud instances is maintained.
- The users of cloud instances are charged according to the services they use, which follows 'pay-per-use' policy of cloud computing.
- The fixed price can be decided depending on the electricity market price of the smart grid. Therefore, the cloud price varies depending on the dynamic price of electricity in the smart grid. Consequently, as discussed in the aforementioned

points, the service providers are not burdened with the dynamic price of electricity for cloud service provisioning.

Similarly, Lucanin and Brandic [14] suggested real-time cloud instance scheduling by considering the electricity market in the smart grid. In such a model, a cloud instance scheduler is introduced to control the instances depending on the real-time energy price. If the real-time energy price is very high, the scheduler pauses few instances (i.e., virtual machines), in order to minimize the operating cost to the service providers. Figure 7.1 presents a schematic view of such a model.

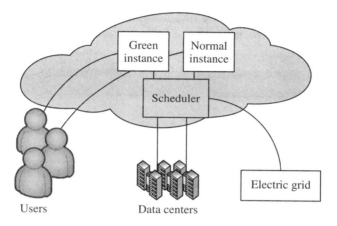

**Figure 7.1** Schematic view of cloud instance scheduler

## 7.5 Future Trends and Issues

- The jobs are scheduled by the EMUs in an adaptive manner in the multi-core processors of the cloud platform. However, the tasks may be executed on different processors placed at different geographically distributed regions. Therefore, it would be productive to take on the technically challenging task of coordinating among the processors at multiple regions while considering the communication cost in the smart grid.

- Current cloud service models follow the 'pay-per-use' concept while still following the static pricing policy for using the cloud services. Therefore, the impact of dynamic pricing in cloud service provisioning can be studied while considering the challenges in *peak* hours similar to the smart grid.

- The cloud instance scheduler schedules virtual machines depending on the real-time energy price in the smart grid. It pauses virtual machines when the price is very high. Consequently, users may not get uninterrupted cloud service, which, in turn, degrades the quality of experience of the users. Therefore, ensuring uninterrupted cloud service provisioning is a challenges, even when the real-time price is high. It can be addressed by introducing the concept of virtual power plant, which will be discussed in Chapter 8.

## 7.6 Summary

In this chapter, different pricing policies were discussed from the perspective of smart grid energy management. Existing distributed pricing policies and their limitations for large-scale deployments were also discussed. Then, cloud-based dynamic pricing schemes, which are applicable to data center networks, were presented. Finally, few future research directions were also presented.

### Test Your Understanding

Q01. What is meant by dynamic pricing?

Q02. What is the difference between static and dynamic pricing?

Q03. What are the challenges of deploying dynamic pricing in smart grid?

Q04. Describe day-ahead pricing policy.

Q05. What is demand-based pricing policy and what is its limitation?

Q06. State the disadvantage of supply-based pricing policy.

Q07. State the drawback of supply–demand-based pricing policy without cloud-based model.

Q08. Describe the problems with the existing pricing approaches without cloud.

Q09. Elaborate cloud-based dynamic pricing policies.

Q10. Point out some future directions of dynamic pricing.

## References

[1] Misra, S., S. Bera, and T. Ojha. 2015. 'D2P: Distributed Dynamic Pricing Policy in Smart Grid for PHEVs Management'. *IEEE Transactions on Parallel and Distributed Systems* 26 (3): 702–712.

[2] Jin, X., Y. K. Kwok, and Y. Yan. 2013. 'A Study of Competitive Cloud Resource Pricing under a Smart Grid Environment'. In *Proc. of the IEEE CloudCom.* pp. 655–662.

[3] Zhou, Z., F. Liu, and Z. Li. 2016. 'Bilateral Electricity Trade Between Smart Grids and Green Datacenters: Pricing Models and Performance Evaluation'. *IEEE Journal on Selected Areas in Communications* 34 (12): 3993–4007.

[4] Sheikhi, A., M. Rayati, S. Bahrami, and A. M. Ranjbar. 2015. 'Demand Side Management in a Group of Smart Energy Hubs as Price Anticipators; The Game Theoretical Approach'. In *Proc. of the IEEE Innovative Smart Grid Technologies Conference (ISGT)*. pp. 1–5.

[5] Wang, P., L. Rao, X. Liu, and Y. Qi. 2012. 'D-Pro: Dynamic Data Center Operations With Demand-Responsive Electricity Prices in Smart Grid'. *IEEE Transactions on Smart Grid* 3 (4): 1743–1754.

[6] Park, J. H., Y. S. Kim, I. K. Eom, and K. Y. Lee. 1993. 'Economic Load Dispatch for Piecewise Quadratic Cost Function using Hopfield Neural Network'. *IEEE Transactions on Power Systems* 8 (3): 1030–1038.

[7] Zhan, Y., M. Ghamkhari, D. Xu, S. Ren, and H. M. Rad. 2016. 'Extending Demand Response to Tenants in Cloud Data Centers via Non-intrusive Workload Flexibility Pricing'. *IEEE Transactions on Smart Grid*. doi: 0.1109/TSG.2016.2628886.

[8] Narayan, A. and S. Rao. 2014. 'Power-Aware Cloud Metering'. *IEEE Transactions on Services Computing* 7 (3): 440–451.

[9] Li, X. and J. C. Lo. 2012. 'Pricing and Peak Aware Scheduling Algorithm for Cloud Computing'. In *Proc. of the IEEE PES Innovative Smart Grid Technologies (ISGT)*. pp. 1–7.

[10] Yamin, H., S. A. Agtash, and M. Shahidehpour. 2004. 'Security-constrained Optimal Generation Scheduling for GENCOs'. *IEEE Transactions on Power Systems* 19 (3): 1365–1372.

[11] Li, Y., D. Chiu, C. Liu, L. T. X. Phan, T. Gill, S. Aggarwal, Z. Zhang, B. T. Loo, D. Maier, and B. McManus. 2013. 'Towards Dynamic Pricing-Based Collaborative Optimizations for Green Data Centers'. In *Proc. of the IEEE International Conference on Data Engineering Workshops (ICDEW)*. pp. 272–278.

[12] Liang, X., X. Li, R. Lu, X. Lin, and X. Shen. 2013. 'UDP: Usage-based Dynamic Pricing with Privacy Preservation for Smart Grid'. *IEEE Transactions on Smart Grid* 4 (1): 141–150.

[13] Erol-Kantarci, M. and H. T. Mouftah. 2011. 'Wireless Sensor Networks for Cost-Efficient Residential Energy Management in the Smart Grid'. *IEEE Transactions on Smart Grid* 2 (2): 314–325.

[14] Lucanin, D. and I. Brandic. 2013. 'Take a Break: Cloud Scheduling Optimized for Real-Time Electricity Pricing'. In *Proc. of the International Conference on Cloud and Green Computing (CGC)*. pp. 113–118.

# CHAPTER 8

# Virtual Power Plant

In a smart grid, distributed energy generators are expected to play an important role in providing cost-effective and green energy-services to customers. Implementation of wind farms and solar panels are two popular distributed energy generating sources. In addition to this, combined heat power (CHP) and fuel cells are also expected to take part in the distributed energy generation process. However, due to the intermittent behavior of such distributed generators, maintaining reliable energy-services to customers is a big challenge for service providers. Additionally, energy generation from the distributed energy sources are environment dependent. Therefore, it may not be cost-effective if more energy is generated during off-peak periods. Consequently, it is required to have some kind of storage devices that can store energy on a situation basis (i.e., during off-peak hours), and can discharge energy to the distribution side during on-peak hours. In order to meet these requirements, a dynamic energy consumption/distribution platform, called virtual power plant (VPP), is considered.

## 8.1 Concept of Virtual Power Plant

A virtual power plant (VPP) is conceptualized as a combination of different distributed energy resources (DERs). Therefore, VPP can be considered a decentralized energy resource system with a large number of small-scale DERs such as solar energy, wind energy, CHPs, fuel cells, and plug-in hybrid electric vehicles (PHEVs). Consequently, it creates a single energy monitoring profile, while considering different parameters characterizing each distributed energy source. Figure 8.1 shows a schematic diagram of a virtual power plant consisting of different energy resources. As a whole, a VPP is capable of providing reliable and cost-effective energy supply to customers. Additionally, a large-scale VPP can be formed with a combination of different small-scale VPPs, as shown in Figure 8.2. Depending on various characteristics, there are two types of VPP – commercial and technical – that will be discussed in this section.

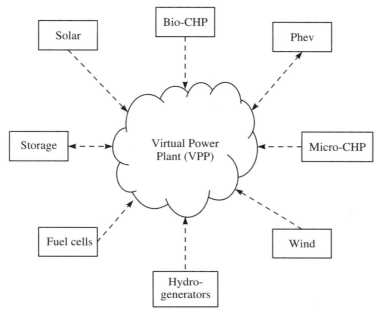

**Figure 8.1**  Schematic diagram of a virtual power plant

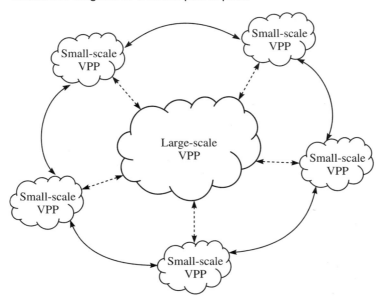

**Figure 8.2**  Large-scale view of virtual power plant

## 8.1.1 Commercial VPP

The concept of a commercial virtual power plant (CVPP) is used for energy trading in energy markets. With the integration of CVPP, energy-market risks can be reduced in order to utilize the distributed energy resources. Consequently, individual DER can benefit from an *intelligent* market in order to maximize its own revenue. A conceptual view of the commercial virtual power plant (CVPP) is shown in Figure 8.3. The operation of a CVPP consists of three phases – input, CVPP operation, and output. Depending on different inputs, CVPP optimizes revenue from individual DERs and offers different services. According to the offered services, each DER schedules its operation time. Additionally, each DER is free to choose any of the available CVPPs, while participating in the energy market. However, CVPPs can define their geographical area for selecting DERs in order to maximize their revenues. Therefore, we can conceive this scenario as a dynamic coalition formation, i.e., a coalition is formed according to the total revenue of the CVPP, while considering individual strengths or capacity. Accordingly, different policies can be incorporated to form the coalition to define the geographical area. Final selection is not done in the CVPP technique.

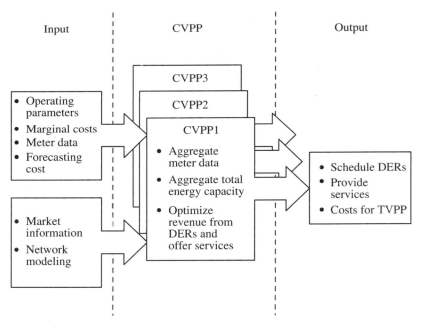

**Figure 8.3** Conceptual view of a commercial virtual power plant (CVPP)

## 8.1.2 Technical VPP

In contrast to the CVPP concept, technical virtual power plant (TVPP) focuses on the visibility of systems operators and system parameters including DERs in its service region. Therefore, TVPP aggregates the capacity and load condition of DERs that have participated in the energy market using the CVPP technique. Consequently, it responds to the characteristics of a DER-system including controllable loads and networks in a single grid's service area, which is known as electro-geographical area. Figure 8.4 shows a conceptual view of the technical VPP. It takes inputs from the CVPP concept in addition to its own parameters and takes decisions based on aggregated load capacity and revenue.

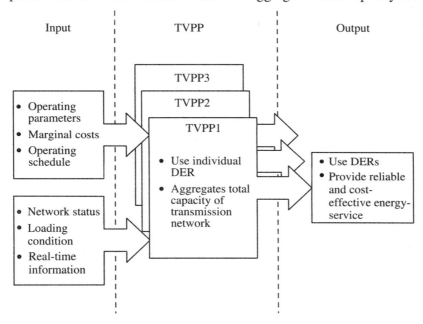

**Figure 8.4** Conceptual view of a technical virtual power plant (TVPP)

Therefore, DERs collect information (such as bidding, geographical position, offers, and other parameters) from different CVPPs. According to the collected information, each DER calculates the visibility of the energy-service using TVPP, while considering real-time or near real-time network management. Finally, the DERs schedule their operating time for which their benefit increases.

## 8.2 Advantages of Virtual Power Plant

### 8.2.1 Acts as internet of energy

Providing electricity to the users as an Internet of energy (IoE) service is one of the important features of the virtual power plant. The IoE concept is used as a real-time interface between smart grid systems and energy-cloud services. Therefore, different renewable and non-renewable energy sources can be integrated together as smart grid components. In the same way, electrical devices, vehicles, distributed energy generators (such as CHP and fuel cells) can be integrated together as energy-cloud services. Consequently, excess amount of energy can be stored in the storage devices (such as electric vehicles and storage devices) during off-peak hours. This stored energy can be used during on-peak hours in order to fulfill the users' energy requirements. Moreover, the users do not need to bother about from where they get the energy-service, which, in turn, establishes the virtual property of the VPP energy-services. Therefore, using the IoE technology, VPP is capable of storing and providing electricity in an efficient manner, so as to maintain a balance between energy supply and demand. Additionally, the VPP architecture of the smart grid will also enable online processing of data received from customers and information sharing through the Internet.

## 8.2.2 Energy efficiency

Energy efficiency is another important advantage of the virtual power plant concept. We have already seen that renewable energy sources are intermittent in nature. Therefore, it is a challenge to predict the exact amount of energy that will be generated from such energy resources. It is a challenge for utility providers to offer reliable energy-services to customers, while there is an uncertainty of energy production from renewable energy sources. When different distributed energy resources (DERs) act individually, most of the DERs do not have sufficient energy resources to fulfill users' requirements. In contrast, in case of VPP, different energy sources and consumers (such as storage devices, electric vehicles) act together to provide a reliable energy management in the smart grid. Therefore, we have a single operating profile using the VPP concept in order to provide efficient and reliable energy-services, while considering the composite parameters of each DER.

## 8.2.3 Online optimization platform

In the smart grid environment, smart meters communicate with the service provider to exchange real-time information (such as energy consumption and price). Therefore, large-scale deployment of the smart grid technology produces large volume of real-time data. Service providers also need to communicate with the generation side for balancing real-time energy supply and demand. Consequently, all these real-time information needs to be processed in order to take adequate decisions. If all the generation and consumption units act in a distributed manner, they also need to have distributed data processing facilities. However, the features of the VPP scheme can deal with such issues and process real-time data from both sides (generation and consumption). As a result, VPP can provide an online optimization platform to offer cost-effective and reliable energy-service to users.

### 8.2.4 Systems security

Using the VPP concept, aggregators (i.e., energy generators) can be monitored through the energy portal. Therefore, utility providers can take adequate decisions to ensure the security parameters in the smart grid. Through the energy portal, energy consumption of the customers can also be monitored, which, in turn, helps to prevent energy theft and smart meters' data manipulation. Additionally, the VPP is capable of handling black-out situations, as different distributed renewable energy sources are operated in a cooperative manner.

## 8.3 Virtual Power Plant Control Strategy

There are three different control strategies for VPP technology – direct, distributed, and hierarchical. In case of direct control strategy, the VPP is controlled in a centralized manner and decisions are also taken by the central unit. In contrast, using distributed strategy, the VPP is controlled in a decentralized manner. In such a strategy, small units take decisions in a distributed manner, and therefore, the service from a particular VPP does not get affected when there is a problem in other VPPs. On the other hand, hierarchical control strategy is a hybrid of direct and distributed control strategies. Therefore, depending on the situation, the hierarchical control strategy can act either in one mode (i.e., direct or distributed) or both. A schematic view of different control strategies for the VPP is shown in Figure 8.5 (adopted from [4]).

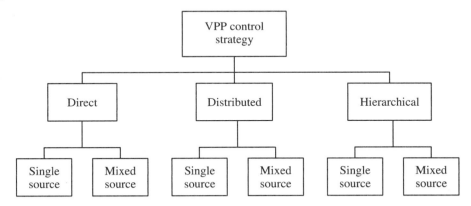

**Figure 8.5** Layered architecture of different control strategies for the VPP

## 8.4 Virtual Power Plant: Different Methodologies

Due to the several advantages of the VPP concept, researchers proposed different VPP methodologies in order to provide energy-service in a reliable and cost-efficient manner. We categorize the existing approaches from different aspects – integration of electric vehicles, implementation of energy storage devices, and other approaches to establish a virtual power plant in a smart grid environment.

### 8.4.1 Integration of electric vehicles

Electric vehicles are gaining interest among power transmission and distribution utilities due to their inherent features–vehicle to grid (V2G) and grid to vehicle (G2V). With the growing concerns about climate change, electric vehicles are expected to play a key role in minimizing $CO_2$ emissions over traditional vehicles. Additionally, the number of electric vehicles is increasing rapidly, which, in turn, is increasing the storage facility in the smart grid. Therefore, implementation of extra dedicated energy storage devices can be avoided for cost-effective energy management in the smart grid. Wind generators are also gaining popularity among researchers for providing clean energy to customers [2]. However, due to the inherent uncertainty in energy production from wind, typically, wind generators are unable to participate in the real-time energy management process for providing reliable energy-services. Therefore, integration of electric vehicles in the smart grid can be an option for reliable energy-services to the customers [2]. Typically, in a day-ahead energy market, power is traded on the $(k-1)^{\text{th}}$ day, so that it can be delivered on the $k^{\text{th}}$ day. Therefore, the VPP system schedules the energy production units according to the energy to be consumed on the $k^{\text{th}}$ day. In order to meet the day-ahead energy trading policy, use of a VPP combined with wind generators and electric vehicles is useful for providing cost-effective and reliable energy service to the customers. Therefore, the objective is to schedule electric vehicles (which are willing to participate) to store excess energy supply from wind generators. Consequently, an optimization model can be formulated for energy exchange between the grid and VPP modules (such as wind generators and electric vehicles). Additionally, a VPP is responsible for optimizing the energy-service from the wind generators. As a result, the VPP controls the estimated energy supply, i.e., it can be either supplied to the grid directly, or it can be stored in the vehicles, or both according to different situations. The optimization problem for the aforementioned strategies can be formulated as follows [2].

$$\underset{\mathbf{G_d},B,\mathbf{G_b},B_c,P}{\text{Maximize}} \quad \mathcal{U}(\mathbf{G_d},\mathbf{G_b}) = \sum_{n=0}^{N-1} p^e(n)[G_d(n) + G_b(n)]$$

subject to

$$G_b(n) + (1+\eta)B(n) + P(n) = Z(n)$$

$$\Delta(n) + B(n) \leq B_c(n)$$

$$\Delta(n) - G_b(n) \geq 0$$

$$P(n) \geq \sigma B_c(n)$$

$$G_d(n) \geq 0, B \geq 0, G_b \geq 0, B_c \geq 0$$

$$0 \leq B_c(n) + P(n) \geq S(n)$$

where $\mathcal{U}(G_d, G_b)$ represents the utility of VPP, while providing electricity in the market. $G_d$, $B$, $G_b$, $B_c$, and $P$ denote the amount of energy supplied to grid, the amount transferred to batteries, the amount transferred from batteries, the available battery capacity, and the energy transferred to vehicles as payments, respectively. The aforementioned parameters form a vector for a particular day and are represented in vector form as $\mathbf{G_d}$, $\mathbf{B}$, $\mathbf{G_b}$, $\mathbf{B_c}$, $\mathbf{P}$, respectively. $\Delta(n)$ is the net amount of energy stored in the vehicle's battery at the start of the $n^{th}$ time-slot, and $S(n)$ is the available storage energy for a particular vehicle. Therefore, the total available storage energy of a VPP system can be represented in a vector form as $\mathbf{S} = \{S(1), S(2), ..., S(n)\}$.

In the optimization problem, the first constraint denotes that the total energy, $Z(n)$, is a combination of energy supplied to the grid, stored energy in the batteries, and energy used for payment. On the other hand, the second constraint guarantees that the electricity stored in the vehicles fits their storage capacities. Similarly, the third constraint guarantees that the discharged energy from the batteries is always less than or equal to the available battery energy. The fourth constraint shows that the received payment to the vehicles is the required proportion of storage energy. Finally, the aggregated energy from vehicles is always less than or equal to the total storage energy. From the solution of the optimization problem [2], we conclude that the total bid in a VPP is the vector sum of the energy directly supplied to the grid from wind generators and from vehicles. Mathematically, it can be represented as $\mathbf{W} = \mathbf{G_d} + \mathbf{G_b}$. We see that the optimization problem mentioned earlier is a linear programming problem. Hence, it can be solved using any linear optimization problem solver.

### 8.4.2 Implementation of energy storage devices

Establishing a VPP makes participation of energy storage devices a distinct possibility. Multiple storage devices can participate in energy trading while considering mutual interests. Recently, an agent-based approach for the management of energy storage devices has been introduced [5]. Figure 8.6 shows an instance of such an agent-based scheme for use in a VPP environment. As shown in the figure, there is a negotiation platform through which multiple agents agree upon a mutual decision. Additionally, multiple agents having deficit or excess energy can co-exist. It is to be noted that a subscriber agent does not know about the source of the energy generator. A seller agent is also unaware of the subscriber. The entire management task is virtualized through the negotiated platform, which is a fundamental concept of cloud computing. Moreover, an adequate balance between energy supply and demand can be established to minimize the

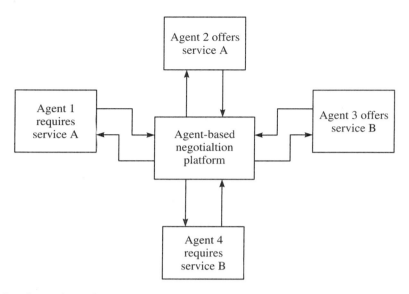

**Figure 8.6** Agent-based energy storage devices management in VPP

waste of excess energy. A practical example is multiple substations participating in a single (negotiated) platform to exchange energy among themselves, while not revealing the identity of each other.

Virtual integration (VI) [6] of distributed energy storage units is useful for effective balancing of energy supply and demand. In such an integrated system, different entities take part in energy management, such as electric store, home, factory, and building energy management system (EMS). All the EMSs are integrated in a common cloud platform through which energy is exchanged among different entities. Figure 8.7 presents such an integrated platform with different EMSs.

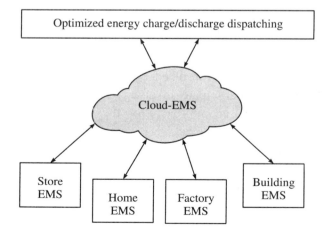

**Figure 8.7** Virtual integration of different energy management units

Through the virtual integration of storage units, dynamic energy charge/discharge, dispatching can be done in an efficient manner, as presented in Figure 8.8.

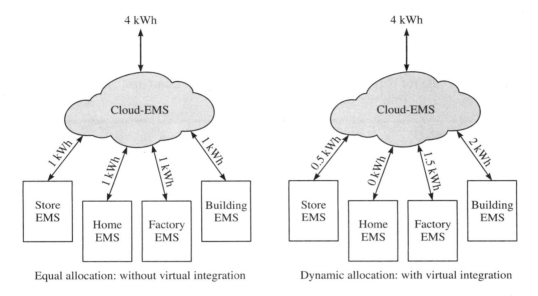

**Figure 8.8** Dynamic energy dispatch through virtual integration

## 8.5 Future Trends and Issues

- It is evident that the integration of multiple devices in a virtualized platform will lead to balancing energy supply–demand in an efficient manner. Therefore, the communication infrastructure plays a big role for large-scale deployment of such systems. The deployment of supporting heterogeneous network infrastructure is a challenging task to successfully establish a VPP, as multiple devices are expected to use multiple communication medium.

- As discussed in earlier chapters, security is also a major concern. In VPP, all the energy storage units are integrated in a common cloud platform. Therefore, privacy-aware resource sharing should be ensured, so that users' privacy is kept secret, as maintained in any virtualized environment.

- Publish–subscribe architecture may be introduced, which would lead to the introduction of service level agreements (SLAs) before charging/discharging the energy storage units. Consequently, dynamic behavior of the VPP system needs to be taken into consideration.

## 8.6 Summary

In this chapter, the integration of different energy units (energy generators and subscribers) to form a virtual power plant (VPP) was discussed. After introducing the basics of VPP, different challenges present in the traditional energy management systems were presented. Then, existing methodologies were discussed to establish a VPP in the smart grid environment in the presence of multiple energy units. It is evident that the VPP will play a key role for real-time energy management in the smart grid.

### Test Your Understanding

Q01. What is a virtual power plant (VPP)?

Q02. What is the need for a VPP?

Q03. What are the different types of VPPs?

Q04. Explain the concept of commercial virtual power plant.

Q05. Explain the concept of technical virtual power plant.

Q06. How does a virtual power plant act as the Internet of energy (IoE)?

Q07. How does a virtual power plant provide energy in an efficient manner in smart grid?

Q08. Explain how a VPP provides an online optimization platform to offer cost-effective and reliable energy-service to users?

Q09. How does the system security provided in the smart grid use the virtual power plant?

Q10. State the different control strategies for VPP technology.

Q11. Explain the different control strategies for VPP technology.

Q12. Describe the different methodologies of virtual power plants.

Q13. State some future trends and issues based on virtual power plants in smart grid.

## References

[1] Pudjianto, D., C. Ramsay and G. Strbac. 2007. 'Virtual Power Plant and System Integration of Distributed Energy Resources'. *IET Renewable Power Generation* 1 (1): 10–16.

[2] Vasirani, M., R. Kota, R. L. G. Cavalcante, S. Ossowski and N. R. Jennings. 2013. 'An Agent Based Approach to Virtual Power Plants of Wind Power Generators and Electric Vehicles'. *IEEE Transactions on Smart Grid* 4 (3): 1314–1322.

[3] Marra, F., D. Sacchetti, A. B. Pedersen, P. B. Andersen, C. Trholt and E. Larsen. 2012. 'Implementation of an Electric Vehicle Test Bed Controlled by a Virtual Power Plant for Contributing to Regulating Power Reserves'. In *Proc of IEEE Power and Energy Society General Meeting.* pp. 1–7.

[4] Raab, A. F., M. Ferdowsi, E. Karfopoulos, I. G. Unda, S. Skarvelis-Kazakos, P. Papadopoulos, E. Abbasi, L. M. Cipcigan, N. Jenkins, N. Hatziargyriou, and K. Strunz. 2011. 'Virtual Power Plant Control Concepts with Electric Vehicles'. In *Proc. of 16th International Conference on Intelligent System Application to Power Systems (ISAP).* pp. 1–6.

[5] Unger, D. and J. M. A. Myrzik. 2013. 'Agent Based Management of Energy Storage Devices within a Virtual Energy Storage'. In *Proc. of the IEEE Energytech.* pp. 1–6.

[6] Sakuma, H., Y. Iwasaki, H. Yano, and Koji Kudo. 2012. 'Virtual Integration Technology of Distributed Energy Storages'. In *Proc. of the IEEE PES International Conference and Exhibition on Innovative Smart Grid Technologies (ISGT Europe).* pp. 1–6.

# Advanced Metering Infrastructure

Advanced metering infrastructure (AMI) is an important component of a smart grid that can help in fulfilling the objectives of the latter. It is a combination of smart meters and bi-directional communication networks. With the help of AMI, customers, service providers, and other entities exchange information among themselves in order to have a well-balanced smart grid environment. Consequently, with the wide deployment of smart meters supported by bi-directional communication networks, it is easier to monitor real-time energy supply–demand information, automatic billing, and many others. Moreover, it is one of the primary requirements for establishing a smart grid environment.

## 9.1 Requirements

Typically, an AMI system should meet the following requirements [1]:

- *Data storage*: Energy consumption data is required to be stored in the smart meter. Therefore, the AMI should help smart meters to store real-time and past energy consumption data. To fulfill this requirement, adequate memory should be installed in the smart meters. Moreover, it is also required to ensure that the stored data is accessible by only authenticated users.

- *Communication*: Different communication protocols are also required to communicate with different entities; examples include the IEEE 802.15.4, IEEE 802.11, and wide area network protocols. For example, home appliances may communicate with the home energy management unit (HEMU) using the IEEE 802.15.4 protocol. On the other hand, a HEMU may communicate with the local aggregator unit using the IEEE 802.11 protocol. Finally, the aggregator units may communicate with the service provider through the protocols used in wide area networks. Therefore, suitable communication protocols are required to be deployed, taking into consideration both their advantages and disadvantages.

- *Management*: Different management functions include remote management, billing management, and load control. All these tasks should be performed in an adaptive manner without the involvement of human beings. In case of remote management, fault in the smart meter must be rectified, and the corresponding action should be taken in order to avoid service disruption. In case of billing management, customers can be automatically billed according to their energy consumption. Finally, in case of load control, energy demand from some of the smart meters can be discarded to establish a balance between the energy supply and demand.

- *User interface*: It can be used as an alternative to users for accessing AMI information; the interface includes LCD displays, graphics constructions, keyboard, and a website with authorization facility.

- *Data reading and processing*: Finally, data reading and processing are important requirements to be fulfilled by the AMI. Based on the processed information, service providers are expected to take decisions. Therefore, the processed information is crucial for optimal management of the system. Otherwise, power outage and unwanted billing can take place in the smart grid systems. Moreover, the grid may fail if there is a large mismatch between the energy consumed through the power line and the energy requested through the communication network.

## 9.2 Different Approaches of AMI

### 9.2.1 Data collection for AMI

Smart meters store energy consumption data at the consumers' end; the stored data is sent to the service providers periodically for the management of the smart grid system. Therefore, different mechanisms are proposed by researchers for efficient data collection from the smart meters, while considering the other requirements discussed in Section 9.1.

The main objectives of an AMI-based smart grid system are the following: low capital expenditure and operational expenditure cost, support of different communication protocols, adaptability of new technologies, and uninterrupted information flows [2]. An entity in the smart grid should be able to communicate with other entities for information exchange. However, this requires universal support of communication protocols. On the other hand, persistent information flows should be ensured even in the presence of adverse environmental conditions, as real-time decisions are taken based on the received information. From this perspective, cognitive radio-based communication infrastructure is useful as it can use different frequency bands depending on the availability of radio links [2]. In such a model, cloud-based information processing and decision making are ideal. The cloud-based platform queries individual smart meters for updated information. Based on the received information, the decision maker at the cloud data center takes appropriate decisions for energy management. Figures 9.1 and 9.2 illustrate the AMI model to collect smart meter data. The model helps in executing energy management in the smart grid.

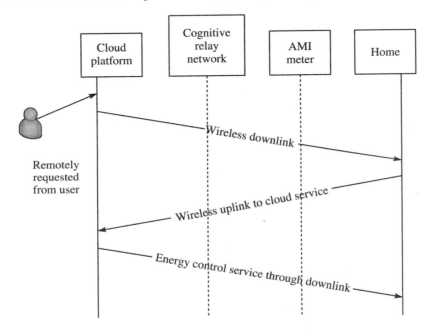

**Figure 9.1** Steps for information collection

Uplink

| RF-HAN protocol | AMI-identifier | Wireless HAN protocol/version | Data |
|---|---|---|---|

Downlink

| Frame type | Time-freq. slots for Tx | Tx RF carrier | Wireless Tx protocol/version | Cloud ref., time | AMI-identifier | Query data from user |
|---|---|---|---|---|---|---|

**Figure 9.2** Header format used for information collection

Figure 9.1 depicts a scenario in which a remote user initiates a query of energy consumption to the cloud-service provider. After getting the request, the cloud-service provider requests the cognitive relay network for access to the available (unused) bands so that the query can be redirected to the AMI system. Consequently, the request is forwarded to the AMI system in the presence of available bands managed by the cognitive relay networks. The AMI system processes the query and gets the required information of the targeted home. Based on the queried data, it replies to the cognitive relay networks. Finally, the information is sent to the remote user. Additionally, the information may be stored at the cloud platform for further processing and computing. It may be noted that the conventional *request-to-send* (RTS) and *clear-to-send* (CTS) signals may be used to get access to the cognitive relay networks. Consequently, in such a system, the unlicensed and licensed bands can be used in the presence of the cognitive relay network, which manages the network dynamically, based on the availability of different frequency bands.

## Data collection through message concatenation

Concatenation is a good technique to aggregate collected data from multiple smart meters. The smart meters' data are concatenated at the data agregator. The main challenge is to aggregate the data while considering the deadlines of individual data. An efficient message concatenation technique has been proposed [3] to aggregate the smart meters' data collected through AMI system in the smart grid. The message concatenation problem is formulated in terms of integer linear programming (ILP), as follows [3]:

Minimize $\quad k = \sum_{i \in n} y_i$

subject to

$$\sum_{j=1}^{n} s_j x_{ij} + H \leq W y_i, \quad \forall i \in \{1, 2, \ldots, n\} \tag{9.1}$$

$$\max a_j x_{ij} \leq d_j x_{ij}, \quad \forall i \in \{1, 2, \ldots, n\}, \forall j \in \{1, 2, \ldots, n\} \tag{9.2}$$

$$\sum_{i=1}^{n} x_{ij} = 1, \quad \forall i \in \{1, 2, \ldots, n\} \tag{9.3}$$

$$y_i \in \{0, 1\}, \quad \forall i \in \{1, 2, \ldots, n\} \tag{9.4}$$

$$x_{ij} \in \{0, 1\} \quad \forall i \in \{1, 2, \ldots, n\}, \forall j \in \{1, 2, \ldots, n\} \tag{9.5}$$

where $k$ is the number of packets to be sent to the data center after concatenation. $y_i = 1$, if packet $i$ is used, and $x_{ij} = 1$, if the message $j$ is put into the message $i$. $s_j$ is the size of the message $j$, $H$ is the fixed header size. Therefore, the total size of the concatenated message including header $H$ should be less then or equal to the maximum allowed size, which is denoted in Equation (9.1). The aforementioned optimization problem can be solved using the linear programming approach by relaxing the integral constraints.

In the traditional smart grid system, the single meter data management system (MDMS) aggregates all the smart meters' data, and sends the aggregated data to the utility center to process it. However, this process poses challenges (specifically overhead) on the MDMS for large-scale deployment, as millions of smart meters are expected to join the energy trading process. Ghasempour [4] analyzed different architectures for an AMI system to aggregate the smart meters' data, with an aim to minimize overhead, while maximizing scalability. Three different architectures were studied – centralized, decentralized, and hybrid – which are discussed here.

*Centralized aggregation*: In such an aggregation scheme, each smart meter is connected to a single data aggregator, which controls the process in a centralized manner. The cost of such an aggregation process can be formulated as an optimization problem as follows:

$$\text{Minimize} \quad C = C_M + \sum_{i=1}^{N_A}(C_i^A + I_i^A), \text{ where } C_M = N_S, N_A = \frac{N_S}{N_P}$$

subject to

$$N_P \leq a, \text{ and } N_A \leq N_F \tag{9.6}$$

where $C_M$ is the cost for MDMS, which is equivalent to the number of smart meters $N_S$, $N_A$ is the number of aggregators, and $C_i^A$ and $I_i^A$ denote the operating cost and installation cost of an aggregator, respectively. The centralized scheme has the following limitations:

- As the smart meters are allowed to connect to a single aggregator, the maximum number of smart meters per aggregator is limited. Therefore, the coverage area may be small, i.e., only the smart meters within the communication range of the aggregator get connected to the service provider.
- Adequate number of substations are required to provide service. If the number of substations are not adequate, the smart meters far from the communication range will not get service.

*Decentralized aggregation*: In case of decentralized aggregation, multiple substations control the data aggregation process and dispatch energy according to the aggregated information. The optimization problem is formulated as follows:

$$\text{Minimize} \quad C = N_S + \sum_{i=1}^{N_U}\sum_{j=1}^{N_A^i}(C_{ij}^A + I_{ij}^A) + C_m$$

subject to

$$N_P \leq a, \text{ and } N_A^i \leq N_F^i (\forall i = 1 \text{ to } N_U), \tag{9.7}$$

$$s_{ij} = 0 \text{ or } 1, \sum_{j=1}^{N_U} s_{ij} = 1, \sum \sum j = 1^{N_U} s_{ij} = N_M, \tag{9.8}$$

$$m_i = 0 \text{ or } 1, \sum_{i=1}^{N_U} m_i = N_M, N_S = \sum_{i=1}^{N_U} N_S^i \tag{9.9}$$

where $N_U$ is the number of substations, $N_A^i$ the number of aggregators connected to the substation $i$ in the AMI system, $N_F^i$ the number of feeders at the substation, and $m_i$ denotes whether the substation $i$ is connected to the utility provider. Finally, $C_m$ denotes the monthly operational cost to the utility provider. Mathematically, $C_m$ is expressed as follows:

$$C_m = \sum_{i=1}^{N_U} \sum_{j=1}^{N_U} (E_{ij}^S \times s_{ij}) + \sum_{i=1}^{N_U} (E_i^U \times m_i) \tag{9.10}$$

*Hybrid aggregation*: In hybrid data aggregation, the advantages of both the centralized and decentralized techniques are harnessed. Therefore, it is possible to have cost-effective data aggregation for an AMI system in the smart grid.

The aforementioned optimization problems can be solved using genetic algorithms [4].

In addition to data aggregation, data routing is an important issue to address in the AMI system. As mentioned earlier, smart meters are resource constrained, which, in turn, means that there is a need to use energy-aware routing algorithms for data forwarding from the smart meters to the utility data center. A low-power lossy-networks-based (RPL) routing algorithm for data forwarding in the AMI architecture was proposed [5] with an aim to maximize network efficiency without increasing the complexity of the system. Rank-based method is applied to forward information from smart meters. The ranks for different possible paths are initiated by the gateway. Figure 9.3 presents an example of a rank-based method. As depicted in Figure 9.3, the path with minimum rank is selected to forward the information.

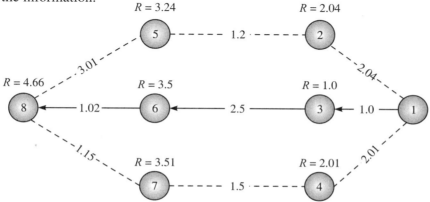

**Figure 9.3** Rank-based routing scheme for AMI architecture

### Data collection through gateways

As discussed in Chapter 1, the gateway aggregates the smart meter information before forwarding it to the meter data management system (MDMS). The IEEE 802.11s protocol can be used as the wireless communication protocol between the smart meter (SM) and the MDMS. Therefore, all SMs are enabled with the IEEE 802.11s protocol and they can create a mesh network with the gateway. Consequently, data from a particular SM can be sent to the gateway through multiple SMs (acting as forwarding nodes). In such a solution, there are two challenges: (a) the position placement of the gateway so that the network performance can be improved while considering the associated cost; (b) the reporting interval of the SMs, as there would be typically a large number of SMs in a smart grid network. Let us focus on these two specific challenges in detail.

*Gateway Placement*: For a given smart grid network, i.e., smart meters with their locations and required reporting intervals (meter reading), the number of gateways required and their placements should be optimized. This problem is NP-hard, i.e., it is not possible to get a complete optimal solution in polynomial time. Therefore, the problem can be solved using some heuristic methods to get a near-optimal solution. One such heuristic scheme was proposed by Saputro and Akkeya [6], who suggested a two-stage approach – in the first stage, a minimum spanning tree (MST) is created using the IEEE 802.11s protocol's path discovery process, and in the second phase, the gateway is selected based on minimum hop criteria. Both of these stages are explained in Figure 9.4.

*Selection of reporting interval*: According to Saputro and Akkeya [6], after creating a minimum spanning tree (MST), the reporting interval can be selected using any of the following three different approaches:

- *Nearest SM first*: In this approach, the SM situated nearest to the gateway is selected first to send its meter data. Then, the second nearest SM is selected, and the process continues. Mathematically, it is determined as follows:

$$T(SM_i) = S_t + (D_i - 1) \times \delta \qquad (9.11)$$

where $T(SM_i)$ is the reporting time for the SM $i$, $S_t$ is a pre-defined constant of all the SMs present in the network, $D_i$ is the depth of the SM $i$ in the MST, and $\delta$ is also a pre-defined constant that adjusts the reporting frequency.

- *Farthest SM first*: In this approach, the SM situated farthest from the gateway is selected first to report its meter data reading. Then, the second farthest one is selected, and the process continues. Mathematically,

$$T(SM_i) = S_t + (D_{max} - D_i) \times \delta \qquad (9.12)$$

where $D_{max}$ is the maximum depth of the MST.

- *Randomized SM selection*: Finally, the SM can also be selected in a randomized manner, i.e., the SMs are chosen randomly to send their meter data reading to the gateway node. Mathematically,

$$T(SM_i) = \text{rand}(1, |SM|) \tag{9.13}$$

where $|SM|$ is the number of smart meters present in the network.

However, these approaches may not be suitable when a large number of SMs is present in the network. In such a situation, the reporting time of each SM can be too large to take an appropriate decision on the service providers.

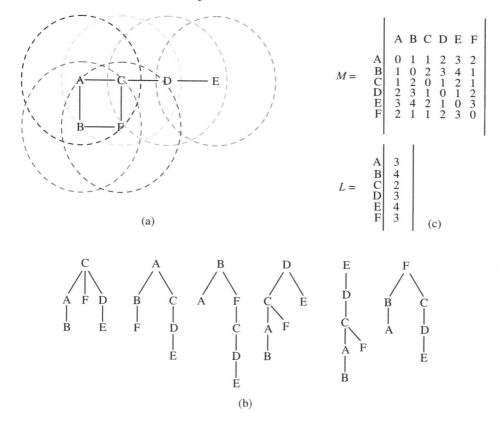

**Figure 9.4** Heuristic-based gateway selection: (a) A sample network topology is created with six smart meters; (b) Possible minimum spanning trees (equal cost for one-hop distance); (c) Obtained minimum distance matrix $M$, and the maximum hop-distance $L$. According to the scheme, C is selected as the gateway node to optimize the hop-distance to all SMs.

## 9.2.2 Classification of AMI data

It is evident that the bi-directional wireless communication network plays a key role in smart grid energy management. The received data at the utility provider is further classified to take coordinated decisions. However, as any wireless network has more data loss compared to a similar wired networks, it may happen that the received data from the smart meters over the AMI system is incomplete in nature, i.e., some of the information is missing. Therefore, it is a challenge to the utility provider to take appropriate decisions in the presence of missing data. To address this issue, a classification approach, which can classify the received data, is useful.

Energy demands from a customer at different time-slots (say 24 slots in a day) are typically highly correlated, i.e., the demand in a particular time-slot is dependent of other time-slots. This is due to the fact that a customer has a fixed energy consumption profile, for example, 5 kWh per day. Therefore, energy consumption in a time-slot will have an impact on the rest of the energy demand, as it has the constraints of total energy consumption in a day. Consequently, the energy demand profile of a customer in the smart grid can be expressed as a multivariate Gaussian distribution [7]. Mathematically,

$$P(l_t; \mu, \sigma) = \frac{1}{l_t} \frac{1}{2\pi |\sigma|^{\frac{d}{2}}} \exp \frac{1}{2} (\ln l_t - \mu)^T \sigma^{-1} (\ln l_t - \mu) \tag{9.14}$$

where $l_t$ is the energy demand profile at a given time $t$, and $\mu$ and $\sigma$ are the mean and covariance of the distribution model, respectively. The time is encoded as $d$ variates (as the model is presented as a multivariate distribution). Therefore, we have $d$-dimensional mean vector $\mu$, which is used to denote the expected load. Further, the system can be modeled as a mixture model with $M$ linear combinations, as follows:

$$P(l_t) = \sum_{m=1}^{M} P(m) P(l_t; \mu_m, \sigma_m) \tag{9.15}$$

This model is learned using the existing data; one distribution, which is likely to be generated, out of the $M$ distributions, is evaluated as a class $c$, as follows:

$$c_t = \max_c \frac{P(c) P(l_t; \mu_c, \sigma_c)}{\sum_{m=1}^{M} P(m) P(l_t; \mu_m, \sigma_m)} \tag{9.16}$$

Typically, the class $c$ can be generated using the traditional approaches such as self-organizing maps and classifiers based on fuzzy logic [8, 11]. However, these approaches require complete data of $l_t$ over all time-slots. In the smart grid communication network, we may not have data at a particular time-slot due to the absence of any energy demand from the customers. As a result, we may need to apply such approaches for classifying the smart grid AMI data when there is missing data.

From the energy demand profile $l_t$ obtained from the customer, a set $s_t$ consisting of all daily energy demands for which the energy demand is valid, can be obtained:

$$s_t = \{a : 1 \leq a \leq A | l_{ta} \in \mathbb{R}\} \tag{9.17}$$

where $A$ is the energy demand profile obtained in advance, and $l_{ta}$ is the advance energy demand profile of $a$. Thus, the dimensions of the multivariate Gaussian distribution is reduced:

$$\mu' \subset \mu, \mu = \{\mu_1, \mu_2, \ldots, \mu_A\} \tag{9.18}$$

Therefore, $\mu'$ can be used with $s_t$ to obtain the daily energy load profile irrespective of the amount of data missing in the AMI system.

### 9.2.3 Security for AMI

In addition to the different routing mechanisms for message forwarding, securing the smart meter data is an important factor. Different security algorithms are introduced to address the security challenges present in an AMI system in the smart grid.

#### Secure communication in AMI

As smart meters are resource constrained in nature, lightweight secure communication would be feasible in the smart grid. A bootstraping protocol is useful to secure smart meter data [10] in the following manner.

1. A smart meter sends its identity with a nonce, which is an arbitrary number used once.
2. After receiving the identity and nonce from the smart meter, the control center sends its identity with a nonce.
3. The smart meter sends its certificate to the control center.
4. After receiving the certificate from the smart meter, the control center verifies it. After successful verification, the control center sends its certificate to the smart meter.
5. After mutual authentication at both ends, the smart meter generates a secret key and its hash. The secret key is encrypted using the public key of the control center. Then, the whole string is encrypted using the private key of the smart meter.
6. The control center sends the hashed data to the smart meter.
7. Finally, the smart meter also sends its data using the hash function.

Figure 9.5 summarizes the steps for securing an AMI system in the smart grid.

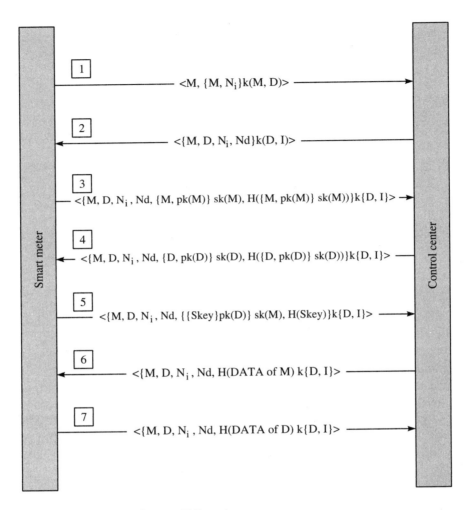

**Figure 9.5** Steps for securing an AMI system

### Intrusion detection in AMI

Several entities are expected to take part in energy trading in the smart grid. Therefore, it is usual to have intruders, who are likely to inject/tamper information, and thus, cause grid failure. Therefore, it is also required to have an intrusion detection system (IDS) for large-scale smart grid deployment, so that intruders can be detected automatically. Towards this objective, Faisal et al. [11] proposed an intelligent IDS system for use with AMI in the smart grid. The major components to facilitate the deployment of IDS are the smart meter, the data aggregator, and the AMI data center.

*IDS at the smart meter level*: The amount of data generated by individual smart meters is relatively very small compared to the data gathered at the aggregator units and AMI

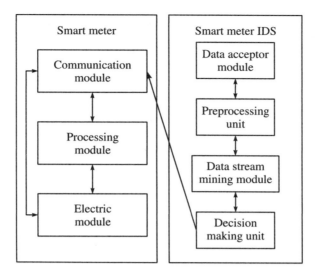

**Figure 9.6** Schematic diagram of a smart meter enabled with IDS

data centers. Therefore, a low data rate will suffice the requirements at the customers' end. To secure smart meters from intruders, IDS can be implemented in the smart meter itself. However, the resource constrained architecture of the smart meter poses challenges in installing IDS. Two solution approaches are feasible to deal with such challenges – (a) a separate IDS for old smart meters, termed as *a security box* [11]; and (b) integrated IDS for new smart meters, as they can be redesigned to support such facility. Figure 9.6 shows the schematic diagram of an IDS-enabled smart metering system. The received information at the smart meter is first processed by the IDS. Then, the information is fetched at the smart meter for other uses. Thus, smart meters can be saved from intruders.

*IDS at the data aggregator level*: Similar to the IDS at the smart meter level, data aggregators can be equipped with IDS. The IDS can be either implemented externally with existing devices or integrated with new devices. Overall, the operation of an IDS at the data aggregator level remains the same. The data aggregators aggregate large data flows received from the smart meters. Further, the information is processed and forwarded to the AMI data center.

*IDS at the AMI data center level*: Finally, the AMI data center collects the global information from all the data aggregators deployed over different regions. The AMI data center is more vulnerable to attacks than the other component, as global information is collected and stored here. An IDS is implemented at the AMI data center end like the IDS at the smart meters, considering the required resources at the data center. Consequently, the entire AMI can be secured by integrating the components with IDS.

### Key management in AMI

Security algorithms use key-based encryption and decryption technologies to ensure data confidentiality. Two types of keys are widely used – *symmetric* and *asymmetric*. In symmetric key cryptography, a shared key is used for both encryption and decryption. In contrast, two different keys are used in asymmetric key algorithms. Efficient key management is also an important factor to ensure data authentication, non-repudiation, and confidentiality. We will discuss different security aspects of smart grid in Chapter 10 in detail. Existing key management techniques used in AMI are discussed here.

A key management framework for AMI-based smart grid was proposed by Liu et al. [12], with the primary objective of securing smart meters. A key graph is formed between the smart meters deployed over a large region in the smart grid. Three types of communication processes are possible – unicast, multicast, and broadcast. Thus, three different key management technologies are also required for message exchange. Moreover, the designed key management technology should be flexible enough to support such requirements in the smart grid. Consequently, a few smart meters can form a group within which a shared key can be used. However, the shared key must be handled in an efficient manner to deal with incoming and outgoing users in the group. Figure 9.7 shows a key management framework based on the key-graph.

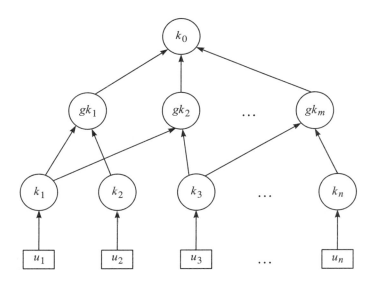

**Figure 9.7** Schematic view of key management framework by forming a key graph

At the top of the tree, $k_0$ denotes the key for global operation in the smart grid. On the other hand, $gk_i$, $i \in N$, denotes the key used for the *i*th group. Individual groups can be maintained in terms of *community* or smart meters connected to a particular service provider. The message size and number of meters within a group is dynamic with time. It may be noted that $k_i$, $i \in N$ and $i \neq 0$, denotes the key for user $u_i$. In such a key-based graph, the key management function (KMF) is the combination of three tuples:

$$\text{KMF} = <\mathcal{U}, \mathcal{K}, \mathcal{R}> \tag{9.19}$$

where $\mathcal{U} = \{u_1, u_2, \ldots, u_n\}$ is the finite and non-empty set of smart meters present in the AMI system, $\mathcal{K} = \{k_1, k_2, \ldots, k_n, gk_1, gk_2, \ldots, gk_m\}$ is the finite non-empty set of keys, in which the subset $\{k_1, k_2, \ldots, k_n\}$ denotes keys for the smart meters, and the subset $\{gk_1, gk_2, \ldots, gk_m\}$ denotes the keys for different groups. Finally, $\mathcal{R}$ is a set containing relations between $\mathcal{U}$ and $\mathcal{K}$. The steps for key generation and key management for unicast, multicast, and broadcast communication are presented here.

*Key generation*: Initially, random keys are generated for smart meters. The initially generated keys are distributed in a secured manner (such as distribution of smart cards). Consequently, the keys ($\{k_1, k_2, \ldots, k_n\}$) and additional values ($\{count_1, count_2, \ldots, count_n\}$) are distributed to the smart meters. $count_i$, $i \in N$, is used to keep track of the number of keys assigned to individual smart meters. The groups are also supplied with keys ($\{gk_1, gk_2, \ldots, gk_m\}$) and count values ($\{gcount_1, gcount_2, \ldots, gcount_m\}$), depending on the number of smart meters and their association with groups.

*Key management for unicast communication*: Typically, unicast communication includes transmission of meter data, joining into or leaving from a group (i.e., service provider), and remote load control. Each time a new session is required to be established used to ensure confidentiality and integrity of the message. The steps [12] for session key generation and usage at the sender and receiver ends, and transmission of the message are described here.

- *Session key generation and usage at sender end*

    1. $C_i = H(HMAC_{k_i}(M_i \oplus CDate) \oplus count_i)$
    2. $sk_i = H(k_i \oplus C_i)$
    3. $EData = E_{sk_i}(Data)$
    4. $Sign_t = HMAC_{sk_i}(EData)$

    Where $H(\cdot)$ denotes the hash operation, HMAC denotes the hash-based massage authentication code, $E(\cdot)$ denotes the encryption operation, and *sign* operation is used to sign a message.

- *Transmission of the signed message*

    1. $u_0(u_i) \to u_i(u_0) : (EData || Sign_t)$

- *Session key generation and usage at receiver end, $u_i$ or $u_0$*

    1. $C_i = H(HMAC_{k_i}(M_i \oplus CDate) \oplus count_i)$
    2. $sk_i = H(k_i \oplus C_i)$
    3. $Sign_r = HMAC_{sk_i}(EData)$
    4. IF $Sign_t = Sign_r$
    5. $Data = DE_{k_i}(EData)$

where $DE(\cdot)$ denotes the decryption operation over an encrypted data.

Therefore, a session is established between the sender and the receiver. The message is signed using the session key, and the signed message is transmitted. The receiver signs the message using the session key and compares it with the received signed message. The receiver accepts the message if both the messages are the same. It is to be noted that for each session, a different session key is generated and used for unicast communication.

*Key management for multicast communication*: As in unicast communication, session key generation and usage at the sender end, message transmission, and verification at the receiver end should be implemented in multicast communication. Multicast communication is useful within a group of users, who want to exchange messages among themselves. All these steps [12] are mathematically presented here.

- *Session key generation and usage at sender end*

    1. $GC_j = H(GCount_j)$
    2. $gsk_j = H(gk_j \oplus GC_j)$
    3. $EData = E_{gsk_j}(Data)$
    4. $Sign_t = HMAC_{gsk_j}(EData)$

- *Message transmission*

    1. $u_0 \rightarrow \{u_i\} : (EData||Sign_t), u_i \in userset(gsk_j)$

- *Session key generation and usage at receiver end*

    1. $GC_j = H(GCount_j)$
    2. $gsk_j = H(gk_j \oplus GC_j)$
    3. $Sign_r = HMAC_{gsk_j}(EData)$
    4. IF $Sign_t = Sign_r$
    5. $Data = DE_{gsk_j}(EData)$

*Key management for broadcast communication*: The steps for broadcast communication are as follows:

- *Session key generation and usage at sender end*
    1. $C_0 = H(Count_0)$
    2. $sk_0 = H(k_0 \oplus C_0)$
    3. $EData = E_{sk_0}(Data)$
    4. $Sign_t = HMAC_{sk_0}(EData)$

- *Message transmission*
    1. $u_0 \rightarrow \{u_1, u_2, \ldots, u_n\} : (EData || Sign_t)$

- *Session key generation and usage at receiver end*
    1. $C_0 = H(Count_0)$
    2. $sk_0 = H(k_0 \oplus C_0)$
    3. $Sign_r = HMAC_{sk_0}(EData)$
    4. IF $Sign_t = Sign_r$
    5. $Data = DE_{sk_0}(EData)$

The aforementioned key management policies are useful to ensure message authentication, integrity, and non-repudiation in a smart grid environment. Therefore, only the intended receiver is able to decrypt the received message. However, the security schemes mentioned earlier are vulnerable to denial-of-service (DoS) attacks [13]. Moreover, a large database and efficient mechanism is required to maintain the keys for multicast communication in the event of large-scale deployment of the AMI system. On the other hand, due to the heterogeneous nature of the AMI in smart grid (i.e., combination of different entities), the existing key management schemes used in traditional networks cannot be used. For use with AMI, a hybrid key management scheme was proposed by Wan et al. [13]. In their scheme, a combination of private and public key cryptography is used to have elliptic curve cryptography. The key management scheme is mathematical presented here.

*End-to-end key establishment*: End-to-end key establishment protocol is used to manage the session keys for unicast and multicast communication. The required steps are as follows:

1. A private key is generated for the smart meter, $M$, by MDMS as: $K_M = sH(M)$, and the public key as: $Q_M = H(M)$.
2. For MDMS, $S$, the private and public keys are as follows: $K_S = sH(S)$ and $Q_S = H(S)$, respectively.
3. The smart meter computes $aQ_M = aH(M)$ based on a randomly generated number $a$.
4. $aQ_M$ is sent to the MDMS.
5. The MDMS computes $bQ_S = bH(S)$ based on a randomly generated number $b$.

6. The MDMS computes a secret session key $K_{SM}^* = H_2(e(K_S, aQ_M + bQ_S)) = H_2(e(H(S), H(M))^{s(a+b)})$.
7. The message authentication code (MAC) key is generated as $k_{SM} = H_3(e(H(S), H(M))^{s(a+b)})$, and the MAC is generated as $MAC_S = HMAC_{k_{SM}}(M, S, aQ_M, bQ_S, 2)$.
8. The MDMS sends $bQ_S$ and $MAC_S$ back to the smart meter $M$.
9. Upon receiving $bQ_S$ and $MAC_S$, $M$ calculates the session key and MAC key as: $K_{MS}^* = H_2(e(K_M, aQ_M + bQ_S)) = H_2(e(H(M), H(S))^{s(a+b)})$, and $k_{MS} = H_3(e(H(M), H(S))^{s(a+b)}$, respectively. Therefore, both the session keys and MACs can be verified easily.
10. $M$ calculates the MAC as $MAC_M = HMAC_{k_{MS}}(M, S, aQ_M, bQ_S, 3)$, which is eventually sent to the MDMS for verification.

$H_2$ and $H_3$ are secure hash functions and $HMAC_k()$ is the hash-based MAC. After generating the session keys at both the ends, it is required to manage the keys.

*Key management for multicast communication*

- *Initialization*: In the initialization process, a pairwise key is generated based on the process of session key generation. The generated key is used as $K_{<l,v>}$ of $M_i$ on the leaf node of the tree. It is calculated as follows:

$$K_{<l,v>} = f(K_{<l+1,2v>}) \oplus f(K_{<l+1,2v+1>}) \tag{9.20}$$

where $f()$ is a one-way function. Therefore, the keys for all nodes in the tree can be calculated. Finally, $K_{<0,0>}$ is the key at the root of the tree for multicast communication.

- *Smart meter joining and leaving*: If a smart meter joins the group, the keys are recalculated and distributed in the same manner. Similarly, if a smart meter leaves a group, the keys are recomputed and allocated at all the remaining smart meters in the group.

The aforementioned scheme is useful for ensuring secure communication in the AMI system of the smart grid. For example, the session keys are computed at the sender and receiver ends based on shared information. Therefore, the secret keys are not shared over the communication channel. Thus, the chances of stealing the secret key are reduced. In a group, all the remaining smart meters are safe from attacks, even if the security of a particular smart meter is compromised.

In public key cryptography, security certificates are issued to its users. The management of certification revocation lists (CRLs) is a challenging task in public key-based cryptosystems. It is expected that millions of smart meters will be integrated in the AMI system to exchange real-time information among themselves. Consequently, a large number of certificates are also expected to be maintained at the certificate issuing authority. Moreover, the size of CRLs may grow over time, as more number of smart meters will participate in energy trading in the smart grid. Typically, the CRLs are maintained in a public server, which should be accessible at all the smart meters in the smart grid for signature verification of other smart meters. Distribution of CRLs for all smart meters will lead to huge communication overheads in the smart grid network. On the other hand, smart meters communicate with local gateways (such as MDMS), which forward the data to the smart grid data center. Therefore, it is not necessary to have certificates of all smart meters at a particular smart meter. Moreover, the resource constrained nature of smart meters should be taken care of, while designing the CRL servers. To address such issues, Akkaya et al. [14] introduced a novel algorithm for the management of CRLs in IEEE 802.11 protocol-based smart grid AMI networks. In such a scheme, multiple CRL servers are designed depending on the locations of the smart meters, instead of having a single CRL server. Certificates can be managed in a group-based manner to reduce the certificate distribution overhead. Figure 9.8 shows a group-based smart grid communication architecture, in which the certificates of smart meters within the group are distributed. From the figure, it is evident that a smart meter is required to have the certificates of other smart meters that are in its group. Consequently, the deployment of cluster-based CRLs is useful for reducing the communication overhead.

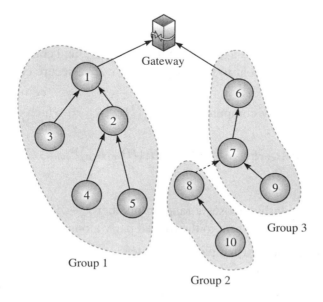

**Figure 9.8** Group-based smart grid communication

**Identity-based security in AMI:** The computational complexity of public key-based cryptography systems is very high. Moreover, these systems may not be suitable for the resource constrained nature of the smart meters. One possibility is to use the identities of smart meters in order to have secure communication in the smart grid AMI network. An identity-based key establishment method is proposed for the AMI system in the smart grid [15]. The proposed key establishment scheme consists of three phases – setup, installation, and key agreement.

**Setup phase:** A trusted party always exists between the smart meter and the AMI control center. A random number $k$ is used as the system parameter. Based on the random number $k$, the trusted party computes the following:

- It chooses a $k$ bit prime number $q$, and constructs $\{F_q, E/F_q, G_q, P\}$. $G_q$ is the set of points on $E/F_q$.
- Then, it computes the public key as $P_{pub} = xP \in E/F_q$ and $x \in Z_q^*$.
- Two hash functions are generated: $H_1 : \{0,1\}^* \times G_q \to Z_q^*$, and $H_2 = \{0,1\}^* Z_q^* \to Z_q^*$.
- Then, it publishes $\{F_q, E/F_q, G_q, P, P_{pub}, H_1, H_2\}$. However, it keeps the $x$ as secret, which is essentially used as the master key.

*Installation phase:* During the installation phase, the trusted authority (TA) does the following using $x$ and the system parameters.

**For AMI control center:**

- The control center chooses a random number $r_{cc} \in Z_q^*$, and computes $R_{cc} = r_{cc}P$. The computed result $R_{cc}$ is sent to the trusted party over a secure channel along with its own id $ID_{cc}$.
- Then, the trusted party computes $y_{cc} = H_1(ID_{cc}, R_{cc})x$, which is sent to the control center.
- The key is valid if $(r_{cc} + y_{cc})P = R_{cc} + H_1(ID_{cc}, R_{cc})P_{pub}$.

**For smart meter:**

- Smart meter $i$ similarly chooses a random number $r_i \in Z_q^*$, and computes $R_i = r_iP$. The computed result $R_i$ is sent to the trusted party along with its id $ID_i$.
- Upon receiving $R_i$, the trusted party computes $y_i = H_2(ID_i, y_{cc})x$, and it is sent back to the meter over the secure channel.
- The smart meter computes $S_i = r_i + y_i$, $S_i$ and $R_i$ are considered as the private and public keys, respectively.

Therefore, at the end of the initialization phase, the smart meter and the control center have all the required information.

*Key agreement phase*: Finally, in the key agreement phase, the smart meter and the control center establish an authenticated session wring the following steps:

- The smart meter chooses another random number $a \in Z_q^*$, and computes $T_i = aP$. A tuple $< T_i, ID_i, R_i >$ is sent to the control center.
- Upon receiving the message from the smart meter, the control center chooses another random number $b \in Z_q^*$, and computes $T_{cc} = b + r_{cc}$, $T_{cc} = T_{cc}P$, and $k_{cc \to i} = T_{cc}(R_i + H_2(ID_i, y_{cc})P_{pub} + T_i)$.
- Then the control center computes $M_1 = H_1(0, k_{cc \to i})$ and sends $T_{cc}$, $ID_{cc}$, and $M_1$ to the smart meter.
- Similarly, the meter computes $k_{i \to cc} = (S_i + a)T_{cc}$, and $M_1' = H_1(0, k_{i \to cc})$. If $M_1 = M_1'$, the authentication process is successful at the meter end. The session key is set as $K = H_1(ID_i || ID_{cc}, k_{i \to cc})$.
- The meter also computes $M_2 = H_1(1, k_{i \to cc})$, and it is sent to the control center.
- Upon receiving $M_2$, the control center calculates $M_2' = H_1(1, k_{cc \to i})$. If $M_2 = M_2'$, the authentication is ensured at the control center end. The session key is selected as $K = H_1(ID_i || ID_{cc}, k_{cc \to i})$.

Finally, the messages are exchanged using the session keys. The session keys defer across sessions in the message exchange process between the smart meter and the control center.

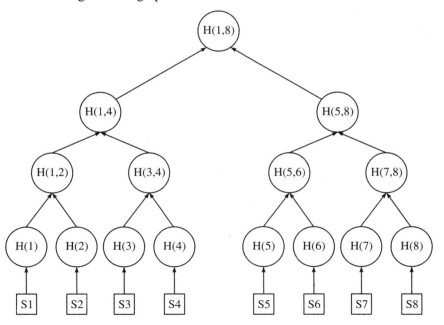

**Figure 9.9** Merkle tree for eight revoked certificates

### Certificate revocation for smart meters

As discussed earlier, smart meters communicate with the service provider while ensuring data authentication and non-repudiation. Consequently, a list of certificates is also expected to be maintained against each of the smart meters. For the large-scale deployment of smart meters, the required number of certificates will be large, which, in turn, poses challenges to maintain such a large database of certificates. Moreover, the certificates of smart meters must be revoked if there is any security breach, such as loss of the certificate. To address this issue, certificate revocation lists can be used to store revoked certificates. There are two options: (a) store all the certificates in one database; (b) store the certificates in a separate database against each of the smart meters. In the first case, the database may be too large to maintain, and in the second case, it may be difficult to handle such a large number of databases. To address this issue, a cluster-based certificate revocation list mechanism was proposed by Rabeih et al. [17]. In the cluster-based scheme, certificates of a group of smart meters are stored in a single database. For certificate revocation, Bloom filter and Markle tree-based approaches are used. Bloom filter is used to store the certificates in a hashed format, so that the searching of a certificate can be done in O(1) time. Figure 9.9 shows a Markle tree for eight revoked certificates.

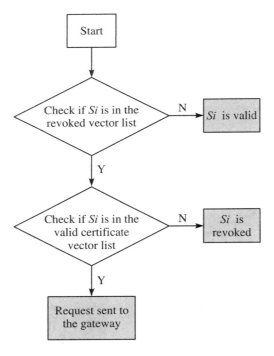

**Figure 9.10** Certificate verification process

From the tree, it is evident that there is a path (called verification path) from the root to the leaf node. Therefore, using the path, it can be verified whether a leaf node belongs to the tree, in which the trees are considered as clusters. Thus, we do not need to create individual databases or even a large database for all the smart meters. Depending on their locations and distributions, multiple trees can be created, and the certificate of a leaf node can be verified. If a leaf node does not exist in any tree, the certificate associated with the leaf node is considered as revoked. In such a case, a new certificate can be issued to the leaf node, whereby it can serve as a member of a suitable tree. The flowchart for the certificate verification process is presented in Figure 9.10 (adopted from [17]). If the certificate is valid after the verification process, communication takes place between the smart meter and the gateway. Figure 9.11 shows the flowchart for verifying whether a certificate is revoked.

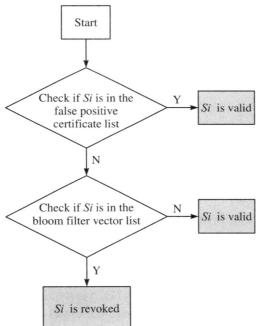

**Figure 9.11** Verification of a certificate to check whether or not it is revoked

## 9.2.4 Electricity theft detection in AMI

In an AMI system, energy theft is another important issue to be considered in order to have reliable and cost-effective energy management in the smart grid. Energy theft is categorized as follows [16]:

- *Physical attack*: As the name suggests, the users physically attack the system, for example, smart meters, and try to minimize energy consumption cost. Such attacks are typically done by tampering with the wire or by disconnecting the meter, thus preventing the meter from reporting the energy usage to the service providers. It is very challenging to the service providers to deal with such physical attacks.

- *Cyber attack*: In such attacks, the data generated from the smart meters are modified, and the wrong (modified) data are sent to the service providers. Therefore, such attacks typically exist in the communication networks.
- *Data attack*: This is a hybrid approach of both physical and cyber attacks. The smart meter values may be changed by physically tampering the meter or by attacking the communication channel between the smart meter and the service provider.

**Theft detection based on usage pattern**

To deal with such attacks, an energy consumption-based theft detection approach is useful [16]. Typically, energy theft occurs due to violation of data integrity in smart grid communication networks, i.e., the actual data is modified before it is received by the intended receiver. The energy consumption-based theft detection approach consists of two phases – training and application phases – which further consist of several sub-phases [16].

*Training phase*: In the training phase, the system is trained in such a way that it is capable of detecting any unwanted activities in the smart grid. It is executed using the following steps:

- There are two types of energy losses in power distribution network – one is transmission loss or technical loss (TL) and another is loss due to energy theft or non-technical loss (NTL). TL is measured based on the physical properties of the wire and transmission distribution lines, while compromising a small error. For example, TL can be due to the transmission loss or power outage due to technical failure in the transmission–distribution lines. Therefore, after approximating the TL, the service provider calculates the NTL from the total power consumption.
- After getting an approximated value of NTL, it is required to process and analyze the data at the service providers' end. Through the cloud platform, the service provider can process and analyze the data in a centralized manner. Smart meters report their energy consumption data to the service provider. Therefore, if a smart meter sends energy consumption data of $n$ instances in an hour, the total number of instances to the service provider is $n \times 24$ in a day. Consequently, for a large-scale deployment of the AMI system, the size of the data will be very large. As a result, it is necessary to implement some dimension reduction algorithm, which can be used to reduce the dimension of the received data. It is to be noted that the interval between two consecutive data from a smart meter varies due to the fact that smart meters can report anytime based on their strategies. Using the dimension reduction method, the data is stored in a suitable format for processing.
- Once we have the data in a suitable format, it needs to be clustered to analyze its distribution patterns. The data can be varied according to the time, day (e.g., weekend or week-day), and environmental factors (e.g., rainy day). Therefore,

$k$-means clustering algorithm is used to classify such data, while considering the factors mentioned earlier. We can have $k$ different distributions that are obtained from the clustered data set.

- Once the data with different distributions are ready, they are used for training purpose. However, it is required to select which sets of data can be used for the training purpose. To train the classifier, existing simple classification approaches can be used. However, in that case, we need to have an accurate data set, which is quite infeasible in a practical scenario in the smart grid. Therefore, a mixture of original samples and some malicious data can be used to train the classifier [16]. As mentioned earlier, the reporting of energy consumption data is inaccurate (less than the original one). Let us assume that the original dataset is presented in a vector form as $\mathbf{X} = \{x_1, x_2, \ldots, x_n\}$ consisting on $n$ samples. Due to the theft factor, the service provider receives this energy consumption data as $\mathbf{Y} = \{y_1, y_2, \ldots, y_n\}$ according to the meter readings. Therefore, for honest customers, $\mathbf{X} = \mathbf{Y}$. On the other hand, it is $\mathbf{Y} = h(\mathbf{X})$ for fraud customers. Moreover, the function $h(\cdot)$ follows a property over the past energy consumption patterns. Thus, it is possible to generate a combined dataset from the original dataset.

- Finally, SVM (or other similar classifiers) can be used to train the classifier. The values of the different parameters used in SVM can be adjusted according to the service providers' requirements. Thus, the system is trained to detect any unwanted behavior observed in the energy consumption reporting phase from the smart meters through the AMI system.

*Application phase*: Once the classifier is trained, application of theft detection is started in the application phase. To detect the theft event, the application phase follows several steps, which are listed as follows.

- The data of smart meters are received at the service provider through aggregators. Therefore, the data received at the service provider's end can be treated as the energy consumption data of a neighborhood. The data received through the communication network is compared with the original consumption through power network, while considering the technical loss and error associated in the TL calculation. If the reported data with TL and error is less than the actual consumed data, then it is inferred that there exists an anomaly in the data.

- To detect the theft scenario, the received data is processed to get the desired format. Then, the data set is applied to the trained classifier. There can be two cases as follows:

    – There is no anomaly in the dataset. Then, the dataset is applied to train the classifier further.

– NTL is detected by the classifier. Then, adequate measures are taken to deal with such a theft scenario.

- There can be another two possibilities – (a) anomaly is not detected in the first step, but SVM detects an anomaly. This may happen due to the misclassification of the training data; (b) the classifier is unable to detect the anomaly although there is an anomaly present in the first step. In this case, adequate signing techniques can be employed to detect such misbehavior.

**Service restoration on theft detection**

Following theft detection, service restoration is required to get rid of such unwanted situations. Similar to the theft detection scheme mentioned earlier, energy theft can also be detected according to the energy usage of the customers and the total loss at the distribution side in a distributed manner [18]. Figure 9.12 shows the detailed flowchart of the theft detection model proposed by Wang et al. [18]. Any consensus-based algorithm [19] can also be used in such a scenario.

## 9.3 Future Trends and Issues

- Licensed bands can be used depending on the availability. Therefore, the cost for communication among the entities in the smart grid can be modeled dynamically like the dynamic price of energy. Moreover, a common interface is required to use licensed and unlicensed bands optimally to get efficient communication network operation in the smart grid.
- The packets can be concatenated when they meet mutual delay requirements. Moreover, the computation time increases in the concatenation process. Additionally, adequate memory and appropriate processors are also required to process the smart meter data received at the aggregator. Therefore, the resource constrained nature of the smart grid communication network needs to be taken into account while employing concatenation of smart meter data.
- In the RPL-based routing, it is evident that a forwarding chain will be formed among the smart meters to forward the smart meters' data. Instead of forming a chain, cluster-based information forwarding may also be useful to have a master–slave architecture in the AMI networks.
- It is important to dynamically change the key size depending on the resource of the smart meter. If the resource is very low, the key size should be reduced to minimize the computation time. Therefore, a dynamic re-sizable key is required in smart grid operation.

Advanced Metering Infrastructure 143

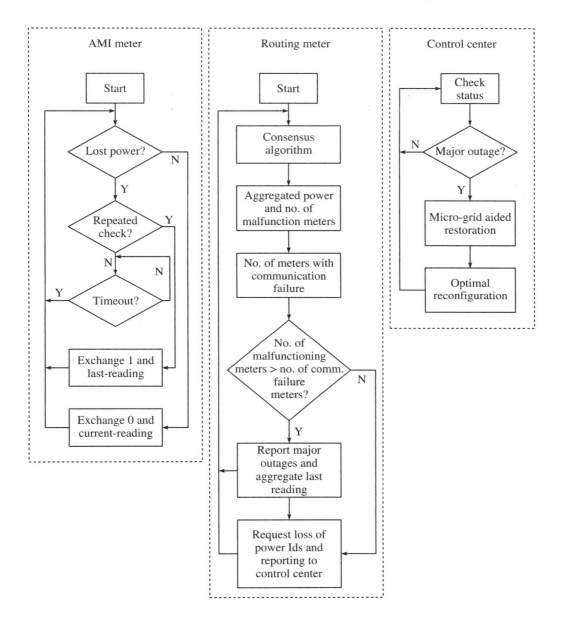

**Figure 9.12** Flowchart for theft detection model

- Traditional security algorithms are used to ensure authentication, integrity, and non-repudiation. Typically, the existing algorithms are power hungry, which also require large volume of resources to compute. Therefore, lightweight security algorithms can be proposed for smart grid applications. For example, lightweight security algorithms are useful to protect the small data generated at the smart meters. On the other hand, traditional heavyweight algorithms can be used at the data centers.

- The proposed CRLs in [14] are applicable in the IEEE 802.11 protocol-based smart grid network. However, smart meters may use different communication protocols for message exchange. To address this issue, the proposed architecture may not be adequate. Moreover, forming groups among smart meters also increases the message overhead, which should also be taken into account.

## 9.4 Summary

In this chapter, different challenges and requirements for deploying advanced metering infrastructure in the smart grid were discussed. Then, the existing AMI technologies, which are useful to address those challenges, were presented from two different perspectives – smart meter data collection through AMI and security in AMI networks. Different security mechanisms to ensure secure communication, intrusion detection, and key management in the AMI-based smart grid system were also discussed. Finally, some of the future research directions were also mentioned based on the limitations of the existing schemes.

### Test Your Understanding

Q01. What is meant by advanced metering infrastructure?
Q02. What are the requirements for an AMI system?
Q03. What are the objectives of an AMI-based smart grid system?
Q04. What are the three different data aggregation architectures?
Q05. Explain the different data aggregation schemes.
Q06. What are the limitations of centralized data aggregation architecture?
Q07. How is the smart meter data secured in AMI?
Q08. Describe the use of intrusion detection systems (IDSs) in AMI.
Q09. Explain the different types of keys used in AMI for encryption and decryption.
Q10. What are the three types of communication processes?
Q11 Mention some future research directions for the AMI system.

# References

[1] Rodriguez, R. H. L. and G. R. H. Cespedes. 2011. 'Challenges of Advanced Metering Infrastructure Implementation in Colombia'. In *Proc. of the IEEE PES Conference on Innovative Smart Grid Technologies (ISGT Latin America)*. pp. 1–7.

[2] Nagothu, K., B. Kelley, M. Jamshidi, and A. Rajaee. 2012. 'Persistent Net-AMI for Microgrid Infrastructure Using Cognitive Radio on Cloud Data Centers'. *IEEE Systems Journal* 6 (1): 4–15.

[3] Karimi, B., V. Namboodiri, M. Jadliwala. 2015. 'Scalable Meter Data Collection in Smart Grids Through Message Concatenation'. *IEEE Transactions on Smart Grid* 6 (4): 1697–1706.

[4] Ghasempour, A. 2015. 'Optimized Scalable Decentralized Hybrid Advanced Metering Infrastructure for Smart Grid'. In *Proc. of the IEEE International Conference on Smart Grid Communications (SmartGridComm)*. pp. 223–228.

[5] Yang, Z., S. Ping, A. Aijaz, and A. H. Aghvami. 2016. 'A Global Optimization-Based Routing Protocol for Cognitive-Radio-Enabled Smart Grid AMI Networks'. *IEEE Systems Journal*. doi:10.1109/JSYST.2016.2580616.

[6] Saputro, N. and K. Akkeya. 2017. 'Investigation of Smart Meter Data Reporting Strategies for Optimized Performance in Smart Grid AMI Networks'. *IEEE Internet of Things Journal*. doi:10.1109/JIOT.2017.2701205.

[7] Harvey, P. R., B. Stephen, and S. Galloway. 2016. 'Classification of AMI Residential Load Profiles in the Presence of Missing Data'. *IEEE Transactions on Smart Grid* 7 (4): 1944–1945.

[8] Seem, J. E. 2005. 'Pattern Recognition Algorithm for Determining Days of the Week with Similar Energy Consumption Profiles'. *Energy and Buildings* 37 (2): 127–139.

[9] Zhang, Y. and A. Arvidsson. 2012. 'Understanding the Characteristics of Cellular Data traffic'. *ACM SIGCOMM Comput. Commun. Rev.* 42 (4): 461–466.

[10] Kumar, V. and M. Hussain. 2014. 'Secure Communication for Advance Metering Infrastructure in Smart Grid'. In *Proc. of the IEEE INDICON*. pp. 1–6.

[11] Faisal, M. A., Z. Aung, J. R. Williams, and A. Sanchez. 2015. 'Data-Stream-Based Intrusion Detection System for Advanced Metering Infrastructure in Smart Grid: A Feasibility Study'. *IEEE Systems Journal* 9 (1): 31–44.

[12] Liu, N., J. Chen, L. Zhu, J. Zhang, and Y. He. 2013. 'A Key Management Scheme for Secure Communications of Advanced Metering Infrastructure in Smart Grid'. *IEEE Transactions on Industrial Electronics* 60 (10): 4746–4756.

[13] Wan, Z., G. Wang, Y. Yang, and S. Shi. 2014. 'SKM: Scalable Key Management for Advanced Metering Infrastructure in Smart Grids'. *IEEE Transactions on Industrial Electronics* 61 (12): 7055–7066.

[14] Akkaya, K., K. Rabieh, M. Mahmoud, and S. Tonyali. 2015. 'Customized Certificate Revocation Lists for IEEE 802.11s-Based Smart Grid AMI Networks'. *IEEE Transactions on Smart Grid* 6 (5): 2366–2374.

[15] Mohammadali, A., M. S. Haghighi, M. H. Tadayon, and A. M. Nodooshan. 2016. 'A Novel Identity-Based Key Establishment Method for Advanced Metering Infrastructure in Smart Grid'. *IEEE Transactions on Smart Grid* doi:10.1109/TSG.2016.2620939.

[16] Jokar, P., N. Arianpoo, V. C. M. Leung. 2016. 'Electricity Theft Detection in AMI Using Customers' Consumption Patterns'. *IEEE Transactions on Smart Grid* 7 (1): 216–226.

[17] Rabieh, K., M. M. E. A. Mahmoud, K. Akkaya, and S. Tonyali. 2017. 'Scalable Certificate Revocation Schemes for Smart Grid AMI Networks Using Bloom Filters'. *IEEE Transactions on Dependable and Secure Computing* 14 (4): 420–432.

[18] Wang, Z. and J. Wang. 2017. 'Service Restoration Based on AMI and Networked MGs under Extreme Weather Events'. *IET Generation, Transmission and Distribution* 11 (2): 401–408.

[19] Garcia, A. D. D., C. N. Hadjicostis, and N. F. Vaidya. 2012. 'Resilient Networked Control of Distributed Energy Resources'. *IEEE Journal on Selected Areas in Communications* 30 (6): 1137–1148.

CHAPTER 10

# Cloud-Based Security and Privacy

Smart grid is essentially a cyber-physical system (CPS) integrated with electric distribution supported with bi-directional communication networks. To support the communication and computing requirements, third party service providers are also expected to participate. Consequently, the third party service providers also access customers' information that is sent through smart meters and backbone networks. As a result, the smart grid system should have appropriate security and privacy policies to secure the smart meters data and other components from unauthorized access. It is a challenging issue to secure the smart grid from cyber attacks in the presence of online connectivity of all components with the Internet [1]. Among different threats that are present in the smart grid in terms of security, energy theft is one of the most important issues to consider. The intruder can access the control point of the smart meters, and thereby, modify real-time information to change energy consumption information reported to utility providers. Therefore, adequate security policies are needed to deal with such issues. In this chapter, we will discuss general security requirements in a communication network. Then specific security issues present in the smart grid and their remedies are discussed.

## 10.1  Security in Data Communication

In data communication, at least two parties (the sender and the receiver) are involved in message exchange. The sender sends a message for an intended receiver. On receiving the message, the receiver takes adequate actions or decisions. In such a scenario, if the sender sends the message directly to the receiver, it may happen that the message is modified or accessed by an attacker. However, the receiver assumes that the received message is the original one sent by the sender. Additionally, it may also happen that an attacker uses the sender's identity to send malicious data to the receiver. Finally, after sending a message, the sender should not be able to repudiate the sent message. Therefore, it is required to implement security mechanisms to prevent such incidents. For example, in banking transactions, appropriate security mechanisms should be in place so that no one other than the actual user can withdraw money. In this chapter, we focus on the cryptography-based security aspects in data communication instead of focusing on other aspects such as intrusion detection and denial-of-service (DoS) attacks. The basic terms used in security are as follows.

- *Encryption*: The process of hiding the original message using some operation is known as encryption. For example, $\{m\}_e = F(m,k)$, where $m$ is the original message and $k$ is the key used for encryption. Message encryption is done at the sender end.
- *Decryption*: After receiving the encrypted message $\{m\}_e$, the receiver should be able to decrypt it to get the original message, i.e., $m = F(\{m\}_e, k)$. Therefore, if a message is encrypted using some method, it should also be decrypted to get back the original message.
- *Cipher text*: The encrypted message is known as cipher text, i.e., plain text is converted to a cipher text to hide the original message. For example, $\{m\}_e$ is the cipher text in which the plain text is $m$.
- *Message digest*: In data communication, it is also required to ensure that the received message is the original one sent by the sender. The message is converted to some other text by applying an operation, so that the original message cannot be retrieved from the obtained text. On receiving the message at the receiver end, the receiver applies the same operation to obtain the digest from the message. Then, the receiver compares both the received and obtained message digests to check the integrity of the received message. The obtained text is known as message digest.

Keys are useful to secure the message in data communication; the message is encrypted with a key (known only to the sender and the receiver) at the sender end. On receiving the message, the receiver decrypts the message by using the key. There exists two types of key-based security mechanisms – symmetric key and asymmetric key-based cryptography.

- *Symmetric key-based cryptography*: As the name suggests, both the parties (sender and receiver) use the same key for encryption and decryption. Therefore, the key is shared between the sender and the receiver. There are several key-sharing algorithms that are used in symmetric key-based cryptography approach. Diffie-Helman key exchange algorithm [5] is one of the popular algorithms used to share the key between the sender and the receiver.

- *Asymmetric key-based cryptography*: In contrast to symmetric key cryptography, different keys are used for encryption and decryption in asymmetric key cryptography. Therefore, each party maintains two keys, which are known as the public key and the private key. The public key, as the name suggests, is known to all. On the other hand, the private key is known only to its owner. In case of encryption, the sender encrypts a message using the receiver's public key. On receiving the encrypted message, the receiver decrypts it using its own private key. Consequently, no other party can decrypt the message as it can only be decrypted using the private key of the intended receiver. Public key infrastructure (PKI) is the backbone of such an asymmetric key-based security approach.

- *Message authentication*: In data communication, the receiver of a message should be able to check whether the received message is sent from an authenticated user or not. If the receiver finds that the message is not sent from an authenticated sender, the message can be discarded to secure the system from any security threat. The process of checking the authenticity of the sender of a message is known as message authentication. For example, a sender A sends a message to a receiver B. After receiving the message, B should be able to check the origin of the message. Otherwise, anyone can send any malicious message to it, which, in turn, can damage the system at receiver B. Through message authentication, the receiver can check the authenticity of the sender.

- *Message integrity*: Message integrity ensures whether the received message and the sent one are the same. This process helps to identify whether the message is modified or not. In smart grid, it is evident that the smart meter data will be forwarded through the backbone network to the service provider. Therefore, the message may be modified by an attacker, which may mislead the service provider. Through the message integrity mechanism, the service provider (i.e., receiver) can check the integrity of the message to ensure that the message is not modified. Before checking the integrity of the message, the hash function to be used should ensure the following.

    - *Pre-image resistant*: The hash function should be pre-image resistant, i.e., one should not get the original message from the message digest. The hash function should be one way. For example, a message digest $H(m)$ is obtained from a message $m$. It is never possible to get back the message $m$ from the message digest $H(m)$. This property of the hash function is called pre-image resistant.
    - *Second pre-image resistant*: This property ensures that for a given message $m$, it is difficult to get another message $m'$, such that $H(m) = H(m')$. Therefore, the hash function should ensure that the same hash function cannot be obtained for two different messages.

– *Collision resistant*: It is similar to the second pre-image resistant, i.e., it is impossible to find two different messages for which the message digest is the same. For example, for two messages $m_1$ and $m_2$, the hash function should ensure the following: $H(m_1) \neq H(m_2)$.

- *Message non-repudiation*: Finally, non-repudiation ensures that a sender cannot repudiate the sent information once the message is sent to a receiver. This is used to ensure that the sender has sent the message and it has been received by the receiver. For example, a customer may deny an energy demand request sent earlier to the service provider. Additionally, the service provider can also repudiate the energy price sent to the customer. Through the message non-repudiation mechanism, both the sender and the receiver cannot repudiate the sent messages. In many cases, non-repudiation at the sender-end is considered. In some cases, it is considered at both the ends. In smart grid system, it is useful to consider message non-repudiation at both the sender and receiver ends, as both energy demand and price are dynamic in nature over time.

### 10.1.1 Overview of PKI

Cryptographic algorithms are of two types – secret key cryptography and public key cryptography (PKC). In secret key cryptography, before deployment, the sender and the receiver share a secret key. The most popular algorithm for key sharing is the Diffie–Hellman algorithm [5]. Secret key cryptography is faster than public key cryptography. However, in secret key cryptography, there is always a chance of leakage of the secret symmetric key at the time of key sharing. Moreover, as the key is shared for one-to-one communication, it becomes unmanageable when the number of users increases. To address this problem, PKC[1] was introduced to ensure message authentication, integrity, and non-repudiation.

In public key cryptography (PKC) [6], each user holds two separate keys, namely, the public key and the private key. The public key of any user is available publicly, but the private key is stored in the user's machine in a very protected manner. For example, if Alice wants to send a message to Bob, Alice uses the public key of Bob to encrypt the message. At the receiver end, Bob uses his own private key to decrypt the message. For encryption and decryption, different types of algorithms such as RSA [7], Rabin cryptography [5], Massey Omura cryptography [5], and elliptic curve [8] are available. Therefore, there is a need to ensure that the public key is associated only with Bob. To address this issue, public key infrastructure (PKI) comes into picture.

A PKI consists of several components.

---

[1] In case of PKC, two separate keys – public and private – are used. Whenever, a sender wants to communicate with a receiver, the former needs the public key of the later. It is required to ensure that the public key used by the sender is the actual public key of the intended receiver. In PKI, the certificate authority (CA) is the trusted third party through which the sender gets the actual public key of the intended receiver.

- Certificate authority (CA): As mentioned earlier, a user maintains a private and a public key. The pubic key is publicly available, and the private key is kept secret. There is a need for an authority to issue such keys and the associated operations used to secure a message in data communication. The authority is called the certificate authority (CA). The CA is responsible for issuing and maintaining the keys. The architecture of CA can be flat or hierarchical. In the hierarchical architecture, multiple sub-CAs are positioned under one root CA.
- Certificate: A certificate consists of a public key and other operations (algorithms) to be used to ensure security. Typically, two certificates issued by the same CA consist of the same operations. For example, if a sender encrypts a message using operation 1, the receiver should also use operation 1 to decrypt the message. Consequently, both the parties follow the same principle for securing the message. Such rules are defined by the CA at the time of issuing a certificate to a user.
- Certificate revocation list (CRL): The CA issues certificates to users. However, if the key of a user is compromised, it should be revoked to avoid any security breach in the system. Such revoked keys are stored in a list known as the certificate revocation list.

Different countries have different specific implementation mechanism PKI.

In India, for example, there is a root certifying authority. The root certifying authority appoints a controller of certifying authority (CCA). Under the CCA, there are several certifying authorities (CAs). CAs issue digital certificates to users. The CCA validates the public key of the CAs to ensure that the certificates provided to users are issued by a licensed CA. CAs are the agents of trust for issuing certificates. The issued certificate has a validity period. The user has a private key corresponding to the public key. A cryptographic token is used to protect the user's private key. If a computer stores the user's private key, then it should have very few connections to the external world. X.509 is a PKI standard for digital certificates [3], which has several attributes, namely, version, serial number, issuer name, validity, subject name, subject public key, and extensions. The subject name or issuer name and serial number act as the key to identify certificates. The main components of the X.509 certificate are end entity, certifying authority, registration authority, certificate revocation list issuer, and repository. When the CA issues a certificate to the user, the certificate is expected to be valid for its entire validity period. However, there are several reasons due to which the certificate becomes invalid before expiration of the validity period. Therefore, the concept of 'certificate revocation' [4] was introduced. The CA maintains a data structure, which is called the certificate revocation list (CRL). The CRL holds the revoked certificate. When a user verifies another user, it not only checks the certificate's validity and CA's signature, but also the recent CRL to know whether the certificate has been revoked or not. The CRL is updated on a regular basis.

In the traditional PKI system, digital signature [15] provides data integrity, data authentication, and data non-repudiation. To retrieve the actual message, the corresponding decryption algorithms are used on the receiver side. To achieve integrity, different hash functions are used to create a message digest. Some of the hash algorithms are SHA256, SHA512, and SHA3 [9]. Integrity is achieved by comparing the message digests created at the sender side and the receiver side from the message. If both the digests are equal, the receiver is ensured that the message is the original one sent by the sender. Non-repudiation is achieved by the use of digital signature and time-stamp. CA is, usually, a trusted third party in PKI. Therefore, PKI is considered as one of the preferred authentication approaches over the user id-password methods. Digital signature is a very important concept in PKI. Millions of people are users of PKI. When HTTP is used for server identity, PKI is used for the SSL handshaking process. Using PKI, one can sign documents digitally. This is done by generating the hash value and then encrypting it with the private key of the sender. PKI is also used for secure email transfer.

There exists several algorithms to secure data communication. However, it is also required to analyze the performance of a particular security algorithm against attacks to the system. This is known as *cryptanalysis*. Some well-known attacks are as follows:

- Cipher text attack [10]: In this type of attack, the attacker has a cipher text and he/she tries to discover the original message. The attacker may also try to obtain the key from the cipher text to get the plain text. If it is possible to get the original message from the cipher text without knowing the key (in advance), the corresponding encryption algorithm is not useful. The algorithm should not be used to secure the system.

- Known plain text attack [11]: The attacker has many plain-texts and corresponding cipher texts. Using both the texts, the attacker tries to find the key, so that the latter can be used to decrypt future cipher texts. Such a type of attack is known as the plain text attack. However, the attacker does not have any access to the sender's system.

- Chosen plain text attack [12]: It is almost similar to the known plain-text attack, but the attacker has access to the sender's system for a temporary time period. In that time period, the attacker chooses few plain texts and the corresponding cipher texts. From both the texts, the attacker tries to find out the key, which can be used in future to attack future cipher texts sent by the sender.

- Chosen cipher text attack [12]: In case of chosen plain text attack, the attacker has access to the sender's system for a limited time period. Sometimes the attacker has access to the receiver's system for a limited time period; this is the chosen cipher text attack. Here, the attacker chooses few cipher texts and modifies those at the receiver's system. Finally, based on the modified cipher texts, the corresponding plain texts are formed. Such an attack is vulnerable to the system, and the underlying security algorithm should not be used.

- Man-in-the-middle attack [13]: Typically, in symmetric key cryptography, a shared secret key is used at both the sender and receiver ends. Therefore, the key is shared in real-time between both the parties. In the key-sharing phase, the attacker accesses the communication channel between the sender and the receiver, and the former modifies the key. The modified key is used at the sender end to encrypt the message. However, the attacker is able to decrypt the message using the modified key. This type of attack is known as the man-in-the-middle attack. For example, man-in-the-middle attack can be visualized in the Diffie-Helman algorithm for key sharing between the sender and the receiver.

- Brute-force attack [14]: As the name suggests, the attacker tries to decrypt the message using all key combinations. This is computationally expensive and cost-expensive as well. Any security algorithm can be attacked using the brute-force attack. Therefore, it is necessary to limit the time period during which a security algorithm will be used. It is not recommended to use the algorithm after the given time period.
- Collision attack [12]: As mentioned before, the message digest is used to compare the original message with that of the received message at the receiver end. Consequently, one should not have the same message digest for two different messages. Hence, the algorithm used to get the message digest should ensure such a property. If the message digests of two different messages are the same, there is a chance of a collision attack. In this kind of attack, it is difficult to prove the originality of the received message. An attacker can change the message without the receiver knowing, as the message digest is the same.
- Replay attack [15]: In this kind of attack, the attacker stores a message, which is communicated between the sender and the receiver earlier. Later on, the attacker sends the stored message repeatedly to the receiver. The receiver assumes that the messages are sent by the sender only. This type of attack degrades the network performance.

## 10.2 Security and Privacy Challenges and Opportunities

There are several challenges from the viewpoint of security and privacy in the smart grid [2]:

- With the increase in the number of components present in the smart grid, it is a challenge for service providers to track the interactions with different entities, while considering the multiple business models.
- The smart grid architecture is more complex than the traditional power system. As a smart grid consists of several sensors, actuators, and support from communication networks, the smart grid communication networks has to be secure. However, this is a difficult challenge to researchers due to the resource constrained nature of such a network.

- Another important challenge is the distributed nature of the smart grid. Distributed generation (DG) units play an important role in real-time energy management. However, DGs may have different policies according to their owners. Consequently, there may not be a common security framework for all DGs present in the network. Therefore, some enhanced security features are required so that this issue can be handled in a centralized manner although they operate in a distributed fashion.
- Finally, utility providers and third party service providers can easily access customers' information. Thus, customers' privacy concern may not be maintained.

Figure 10.1 shows different security aspects of the smart grid, which need to be addressed.

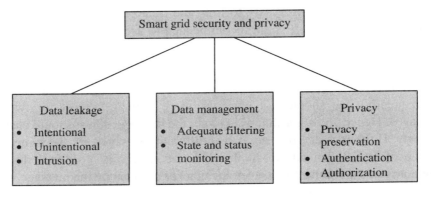

**Figure 10.1** Different security aspects of smart grid

- Data leakage: As discussed earlier, data leakage is one of the most important issues to consider from the security perspective of smart grid. It can be categorized as *intentional*, *unintentional*, and *intrusion*-based. In intentional data leakage, customers can intentionally change something in their smart meters, so that energy consumption cost is minimized. Similarly, utility providers can also change data to over-bill customers. Therefore, adequate security mechanisms should be deployed to address such issues. Similarly, unintentional data leakage can happen due to the failure of any component in the smart grid. For example, due to the failure in the communication network, data outage can happen. Finally, the presence of intruders in the smart grid system can also cause data leakage.
- Data management: Real-time information is generated from different sources. Hence, it is evident that multiple parties may access different types of data for real-time management in the smart grid. Consequently, appropriate access policies are required to be defined so that only the authorized users can access the data. To deal with this issue, suitable data forwarding mechanisms are required.
- Privacy preservation: Finally, it is also expected that a smart meter can forward other smart meters' data. As a result, the forwarding smart meter may access and

process the data coming from other meters before the actual forwarding is done. In such cases, the customers' information may be revealed to others, which is completely unexpected from the viewpoint of privacy. Therefore, ensuring privacy is a big challenge in smart grid.

To deal with the aforementioned issues, a promising approach is to integrate cloud computing applications in the smart grid; cloud computing has many inherent features such as centralized control and provisioning of a common platform to have homogeneous policies.

## 10.3 Security and Privacy Approaches without Cloud

Before discussing the cloud-based security and privacy approaches, let us focus on some of the existing security and privacy schemes proposed in the literature to ensure authentication, integrity, and non-repudiation. It is to be noted that such schemes can also be applied to secure smart grid communication systems.

### User authentication

Smart card can be used to authenticate smart grid users to access any system such as reading smart meter data, participating in energy trading, and so on. Researchers proposed different schemes to ensure user authentication using smart cards. A brief overview of the existing schemes are presented here. Hwang and Li [16] proposed a smart card-based authentication scheme to authenticate remote users. In such a system, it is not required to maintain a password table; the password can be changed anytime. The proposed scheme is also able to prevent message replaying attacks. Further, Hwang et al. [17] proposed a smart card-based simple authentication scheme for remote users. The proposed scheme has four-fold advantages – a password is not required to log-in, message replay attack can be prevented, users can arbitrarily change their passwords, and the password is not stored in the server. The proposed system is more secure compared to other smart card-based authentication systems as the password is not stored in the server. Similarly, Das et al. [18] proposed a smart card-based user authentication scheme in which the ID of the user is used to authenticate the systems. The ID of the user changes over time, which, in turn, secures the user's ID from theft. Additionally, the proposed scheme also resists different attacks such as reply attacks, forgery attacks, guessing attacks, insider attacks, and stolen verifier attacks [18].

Liao et al. [19] proposed a password-based authentication scheme to authenticate users. The proposed scheme deals with different security issues, while maintaining different requirements – the password in not stored in a table; it can be changed freely; it cannot be revealed by the administrator at the server end; it is not transmitted in plain text format on the network; and the user can authenticate the server in addition to the authentication of himself/herself, to name a few. The one-way hash function and the discrete logarithm problem are used to authenticate users. Consequently, the Diffie–Hellman algorithm is used to derive the session key in the proposed scheme.

Luca et al. [20] proposed a user authentication scheme based on touch-screen patterns. In this scheme, the user stores a pattern to unlock the device according to his/her choices. The device recognizes the user input and the way the user performs input. Accordingly, the user is authenticated if and only if both the user input and the performed way of input are matched with the stored one. This authentication scheme is useful for current Android devices and other devices that support touch-screen patterns.

Sae-Bae et al. [21] proposed a multi-touch gesture-based user authentication scheme. In this scheme, biometric techniques with gesture inputs are combined with multi-touch surface to produce an authentication policy. The users authenticate themselves to a system, while providing adequate gestures on the multi-touch surface. A classifier is built to detect the uniqueness of the gestures provided by the users. The authors showed that the classifier can detect the gestures with 90% accuracy. Further, users' ratings are collected to improve the accuracy of the classifier, which, in turn, improves the performance of the proposed scheme. However, the performance of the proposed system solely depends on individual users. Such user authentication algorithms can be used to have authorized access to different components in the smart grid system.

## Message authentication

As we know, a smart grid consists of several entities, which communicate among themselves for cost-effective energy management. Therefore, it is also required to ensure that the messages received by an entity is actually generated from an authenticated entity. Similar to user authentication, researchers proposed several schemes to ensure message authentication. We give an overview of some of the popular schemes. Perrig et al. [22, 23] proposed two authentication schemes for multi-cast streams over lossy channels. The authors proposed two schemes – time efficient stream loss-tolerant authentication (TESLA) and efficient multi-chained stream signature (EMSS). TESLA provides sender authentication, low overhead, and high scalability. However, as mentioned by the authors, it also incurs some delay and losses in initial time synchronization. EMSS provides non-repudiation of source and low overhead. However, like TESLA, it incurs some delay for verification.

Refaei et al. [24] recommended a data and source authentication mechanism, which is mainly based on hashing. Hash chaining is explicitly employed to ensure that no unauthorized modification, insertion, and deletion are done on the data. They discussed the computation, communication, and storage overhead present in the existing authentication schemes that are useful to the named data networking (NDN) infrastructure. On the other hand, in the algorithm proposed by Refaei et al. [24], the computation overhead is reduced; it is quite lightweight compared to standard NDN.

Agren et al. [25] discussed hardware-aligned message authentication based on universal hash functions. They used Toeplitz matrices (a descending diagonal matrix where each left to right descending diagonal element is constant) [28] for authentication.

## Integrity

Integrity is also an important aspect to consider in the smart grid system; it is important to ensure that the message received by a receiver is the actual one generated from the sender. This is important in case of real-time energy consumption, energy generation, and pricing reporting. We also highlight some of the integrity schemes proposed in the recent past. Additionally, the integrity schemes that are useful in a cloud-based system are discussed. Wang et al. [29] proposed a mechanism for ensuring integrity of digital objects, which are stored for a long time. In this scheme, one-way hashing is combined with a time-stamping certificate to verify integrity. In this scheme, a time-stamping certificate is generated for each asset each time. The time-stamping certificate holds a signature, which is generated by the time-stamping authority (TSA). The TSA generates a signature with its own secret key. Using this time-stamping certificate, both the certificate's integrity and the content's integrity are checked. However there is a chance of TSA's secret key leakage. Thus, through the process, one document's hash is combined with another document's hash, termed as hash linking [29], to record the modification of the content.

Yin et al. [30] proposed a secure data storage mechanism at the cloud. The sender encrypts the data before storing at the cloud. After downloading it from the cloud server, a client decrypts it with its own private key. The other users get the public key of the sender from the certificate list and decrypt the message with the public key. One validity period is appended to the certificate. However, if the private key is compromised, the users should be removed from the system, as the public key is an arbitrary string. As a solution, users have to update their private key periodically by contacting the private key generator. Therefore, a secure channel is needed, and if the number of users grows, it will become an overhead. A key update cloud-service provider (KU-CSP) was proposed by Li et al. [3]. Only some of the operations are done by PKG; the others are omitted. After generating the key, the PKG becomes offline. There is another service provider for the key update mechanism. PKG sends the revocation lists to KU-CSP [3].

Tseng et al. [32] identified two drawbacks of using KU-CSP. First, using KU-CSP, computation and communication costs become higher, and second, KU-CSP should hold a secret key for each user, which is an additional overhead. Tseng et al. [32] proposed a cloud revocation authority (CRA) to solve these shortcomings. CRA holds one system secret for all users. One master time key is used, which updates the key periodically. In the cloud environment, sharing data with a large number of users considers efficiency, data integrity, and data privacy. One solution is ID-based ring signatures. Wang et al. [33] proposed the identity-based remote data possession checking (IB-RDPC) protocol, in which there is no need for certificate validation. To ensure the integrity of cloud data, Wang et al. [34]

proposed a security mediator to generate the signature of cloud data. A third party auditor is proposed by Wang et al. [35] to audit the data storage system and guarantee that the cloud data is not modified.

## Non-Repudiation

In addition to authentication and integrity, non-repudiation is another important aspect to consider, in order to resist repudiated action and replay attacks. Again, we briefly discuss some existing non-repudiation schemes, which are useful in smart grid systems. Zhou et al. [36] proposed a protocol for fair non-repudiation of origin and receipt, where both the end users have equal advantage at all times. This protocol does not make use of the simultaneous exchange of secret data in between messages, as its fairness depends on the assumption that the two users are of equal computational abilities. In such a scheme, a trusted third party (TTP) is used. The message is divided into two parts, a *commitment* and a *key*, which are linked by a unique label. The commitment part is sent to the receiver, while the key part is sent to the TTP. Thus, in case a dispute arises, both the users can retrieve the key from the TTP as evidence. Another important aspect of this protocol is that the TTP deals only with the keys, not the message, and hence, TTP's involvement to ensure non-repudiation is much reduced. However, this protocol suffers from two main drawbacks. First, it assumes that the entries made by the TTP are valid and are accessible publicly. Second, keys used in the protocol are associated with the message, which, in turn, may need the help of labels to be defined. Thus, a logic extension is required. The possibility of a replay attack on the protocol suggested by Zhou and Gollmann [36] was explored by Muntean et al. [37] and the necessary modifications to prevent such a situation was also proposed. According to the authors, data freshness needs to be ensured in order to prevent replay attacks. Time-stamps and nonces are suggested by the authors as possible solutions to ensure freshness of data. Due to certain drawbacks of time-stamps such as the need for synchronization, authentic clocks and possibility of window attacks, nonces are used by the authors in this scheme.

Zhou et al. [38] developed the protocol for ensuring non-repudiation, which was theoretically proposed by Zhou and D. Gollmann [36], by reducing the role of the trusted third party further. In this protocol, the users take help from the TTP only when they cannot resolve a dispute between them on their own. This protocol relies on the fact that the parties mostly remain fair to one another and the communication channel between them is never broken. The issue with the label is solved by never letting the originating user select a label, which is not dependent on the message. Additionally, a one-way hash function is used for the unique interpretation of the message. However, the scheme suffers from the existence of a single point failure, as only a single TTP is used. Kremer et al. [39] proposed the use of game theory as a new approach towards the solution of the non-repudiation problem. According to the authors, visualizing the issue of non-repudiation as a game is more helpful in modeling the various aspects of the multi-party communication scenario. The sender, receiver, TTP, channel, and the attacker are all modeled as players of the game. The steps of the protocol are not fixed, and thus, are arbitrarily checked for all possible permutations. The properties of the game are modeled as strategies. The game theory approach may be a sort of turning point in the visualization of the problem of non-repudiation. However, the complexity of the proposed algorithm is very high. Forming the strategies for different scenarios and properties is also another challenge.

Carroll [40] invented a system to ensure complete non-repudiation by making use of several session keys and time-stamping. The structure followed is mainly based on the framework of public key infrastructure. The main advantages of this proposed system is that, by making calculations, it can be proved that the message is sent and received by the sender and receiver respectively, without the cooperation of one another. Whether the information was corrupted or not can also be proved. This method uses an inline TTP or server to control what happens to the 'data parcel' sent by the sender to the receiver. The parcel to be sent is encrypted by several layers of encryption using the sender's private key and receiver's public key. Hash is created from the data file for authentication purpose; a session key is generated and encrypted for validation purpose. A time-stamp is also provided by the TTP server.

Chen et al. [41] suggested certificateless signature (CLS) schemes for non-repudiation. Digital signatures consist of two parts – one done by the signer and the other done by the verifier. The signer creates the signature key, while the verifier verifies the key generated. Identity-based public key cryptography (IB-PKC) and certificateless public key cryptosystems (C-PKC) were traditionally designed protocols where TTP was not required. The CLS strategy works in three steps, which are used in initial setup, signature generation, and its verification. Initially, during the signature setup phase, the user provides a private key as input and the public key as output. Thereafter, the signature is generated, and finally verified against any malicious identity. The results of the different signature schemes (IB-PKC, C-PKC) were compared. The better the level of trust provided by TTP, the better are the security levels. This protocol furnishes non-repudiation of origin, where a user has only one key pair. Eventually, it is claimed that the signature verified by the CLS scheme assures strong non-repudiation.

## 10.4 Cloud-Based Security and Privacy Approaches

There are several security schemes that are designed to secure smart grid communication. We will discuss cloud-based security policies from the perspective of message authentication, integrity, and non-repudiation.

### 10.4.1 Identity-based encryption

The use of the identity of the users is a promising approach to ensure message authentication in data communication. A similar concept is also used in smart grid communication to secure the smart meter data. A cloud-based and big-data driven security framework for data communication in the smart grid was proposed [42]. Figure 10.2 presents a schematic view of the proposed framework. All the entities in the smart grid communicate with one another through the centralized cloud platform. The secure data communication between local servers and the cloud is established using public key cryptography. In public key cryptography (PKC), message authentication and non-repudiation are ensured using two different keys. The user has a public key and a private key. The private key is used to sign a message before sending it to a receiver. The receiver checks the authenticity of the message by using the public key of the sender. Therefore, the public key of a user may be known to all, but the private key remains secret to the user.

Typically, the smart meter data can be forwarded to the utility provider using both the private and the public networks (e.g. the Internet). Therefore, the received information may be vulnerable to the utility provider. Table 10.1 shows the security features that need to be ensured for data communication in the smart grid.

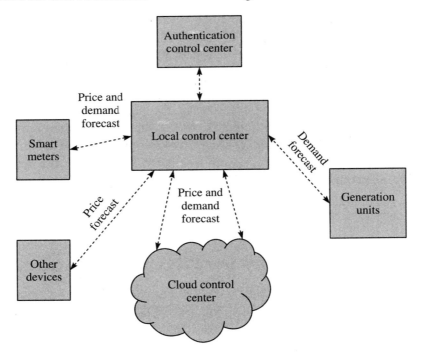

**Figure 10.2** Schematic view of cloud-based secure data communication in smart grid

**Table 10.1** *Security features in data communication in smart grid*

| Purpose | Confidentiality | Data integrity | Non-repudiation |
|---|---|---|---|
| Price forecast | – | ✓ | ✓ |
| Energy forecast | ✓ | ✓ | ✓ |
| Raw energy forecast | ✓ | ✓ | ✓ |
| Monitoring data | ✓ | ✓ | ✓ |
| Control message | ✓ | ✓ | ✓ |
| Other information |  | ✓ |  |

To secure such data communication in the smart grid, an identity-based encryption scheme was also introduced by Ye et al. [42]. Typically, ID-based security schemes are based on public key cryptography. The main advantages of the ID-based security scheme are as follows: (1) ID of the user, which is unique, is used to generate a public key; (2) the key can be computed in advance as well. Therefore, the computation time in real-time can be reduced. Before describing the ID-based security algorithms, let us focus on some of the properties related to this security scheme. Let $G_1$ and $G_2$ be the groups of prime order $p$. Let $g$ be the generator of $G_1$ and $\hat{e} : G_1 \times G_2$. Then, $(G_1, G_2)$ are the bilinear map groups if $\hat{e}$ has the following properties:

- *Bilinearity*: $\hat{e}(aP, bQ) = \hat{e}(P, Q)^{ab}, \forall P, \forall Q \in G_1$ and $\forall a, \forall b \in \mathbb{Z}_p$.
- *Non-degeneracy*: $\forall P \in G_1, \hat{e}(P, Q) \neq 1, \forall Q \in G_1 \setminus \{O\}$.
- *Computability*: There always exists a polynomial time algorithm for computing $\hat{e}(P, Q), \forall P, \forall Q \in G_1$.

Now, we will discuss the ID-based encryption method, which consists of five phases – setup, key generation, signature-based encryption, decryption, and verification [42].

**Setup phase**
In the setup phase, the public key generator (PKG) chooses the following:

1. A group $(G_1, G_2)$ with prime order $p$.
2. A generator $g$ of $G_1$.
3. A random master key $k \in \mathbb{Z}^*$.
4. A domain secret $g_1 = kg \in G_1$.

5. Three cryptographic hash functions: $H_1 : \{0,1\}^* \to G_1$; $H_2 : \{0,1\}^* \to \mathbb{Z}_p^*$; and $H_3 : \{0,1\}^* \to \{0,1\}^n$.
6. The domain public parameters are defined as $P_{\text{param}} = <G_1, G_2, g, g_1, p, H_1, H_2, H_3, n>$.

### Key generation

After the setup phase, key is generated, which will be used for encryption. For a given ID and time-stamp $t$, the key generation algorithm produces a public key, $pub_{ID}$, and a private key, $pri_{ID}$, as follows:

1. $pub_{ID} = H_1(ID||t)$
2. $pri_{ID} = kpub_{ID}$

Therefore, the time is concatenated with $ID$, so that a unique key is generated at each time period. Moreover, key revocation for users who left the system can be done easily.

### Signature-based encryption

For encryption, a sender $A$ encrypts a message $M$ using the following steps:

1. A random number $r \in \mathbb{Z}_p^*$ is chosen by $A$, and it computes $U = rg \in G_1$ and $h_1 = H_2(M||A||U) \in \mathbb{Z}_1^*$.
2. The sender $A$ sets a parameter $V$ as: $V = pri_A h_1 + rg_1 \in G_1$.
3. The public key of the receiver $B$ is chosen as: $pub_B = H_1(B||time)$; other parameters are calculated as: $h_2 = H_2(A||B) \in \mathbb{Z}_1^*$ and $X = h_2 U \in G_1$.
4. It computes $h_3 = H_3(X||\hat{e}(rg_1, h_2 pub_B))$.
5. It sets $W = M \oplus h_3$.
6. Final encrypted output is sent as $<U,V,W,X>$.

In the four-tuple of the encrypted output, i.e., $<U,V,W,X>$, $<U,V>$ and $<W,X>$ represent digital signature and cipher text, respectively.

### Decryption

Once the receiver $B$ receives the encrypted message sent from the sender $A$, $B$ decrypts the message using the following steps:

1. $B$ computes $h_3' = H_3(X||\hat{e}(X, pri_B))$.
2. Then, it decrypts this as $M = W \oplus h_3'$.

Therefore, using the decryption method, the receiver gets the message $M$. However, it should be verified by $B$; the digital signature of the sender has also to be checked, which is presented in the next section.

### Verification
The steps for digital signature verification are as follows:
1. $B$ computes $pub_A = H_1(A||t)$, and $h_1 = H_2(M||A||U)$.
2. The digital signature is verified as original if $\hat{e}(g,V) = \hat{e}(g_1, pub_A h_1)\hat{e}(g_1, U)$.

It is to be noted that the signature, encryption, and decryption can be done depending on the requirements of the applications. For example, if only integrity is the main concern in data communication, we can avoid the use of any encryption method. In such a case, only hashing will suffice the requirement.

## 10.5 Future Trends and Issues

- Smart meters have very limited computation power. Therefore, executing complex security algorithms in the smart meter poses challenges to researchers. Although different security mechanisms are discussed in the literature, there is no clear road-map for securing smart meters and the smart meter data from unattended access.
- Due to the growing interests of smart energy management, the number of smart meters is also expected to increase. Therefore, managing the smart meter data in a secure manner is an important issue in the smart grid.

## 10.6 Summary

In this chapter, different cloud-based security mechanisms were discussed. These are useful to ensure data authentication, integrity, and non-repudiation in the smart grid. After discussing different security challenges involved in smart grid communication networks, an identity-based encryption mechanism was discussed. Finally, the chapter concluded by highlighting some of the future research trends and issues in the context of smart grid security.

### Test Your Understanding

Q01. What are the security and privacy challenges present in the smart grid?
Q02. State the different security aspects of smart grid.
Q03. Explain the different cloud-based security and privacy approaches.
Q04. What is meant by identity-based encryption?
Q05. State the five phases of the ID-based encryption method.
Q06. State some future trends and issues on security and privacy in smart grid.

# References

[1] Liu, J., Y. Xiao, S. Li, W. Liang, and C. L. P. Chen. 2012. 'Cyber Security and Privacy Issues in Smart Grids'. *IEEE Communications Surveys and Tutorials* 14 (4): 981–997.

[2] Bera, S., S. Misra, and Joel J. P. C. Rodrigues. 2015. 'Cloud Computing Applications for Smart Grid: A Survey'. *IEEE Transactions on Parallel and Distributed Systems* 26 (5): 1477–1494.

[3] Zhu, W. T. and J. Lin. 2016. 'Generating Correlated Digital Certificates: Framework and Applications'. *IEEE Transactions on Information Forensics and Security* 11 (6): 1117–1127.

[4] Ganan, C., J. Mata-Diaz, J. L. Munoz, J. Hernandez-Serrano, O. Esparza, and J. Alins. 2012. 'A Modeling of Certificate Revocation and Its Application to Synthesis of Revocation Traces'. *IEEE Transactions on Information Forensics and Security* 7 (6): 1673–1686.

[5] Perlman, R. 1999. 'An Overview of PKI Trust Models'. *IEEE Network* 13 (6): 38–43.

[6] Huang, X., J. K. Liu, S. Tang, Y. Xiang, K. Liang, L. Xu, and J. Zhou. 2015. 'Cost-Effective Authentic and Anonymous Data Sharing with Forward Security'. *IEEE Transactions on Computers* 64 (4): 971–983.

[7] Raj, S. P. and A. P. Renold. 2015. 'An Enhanced Elliptic Curve Algorithm for Secured Data Transmission in Wireless Sensor Network'. In *Proc. of Global Conference on Communication Technologies (GCCT)*. pp. 891–896.

[8] He, J., H. Chen, and H. Huang. 2010. 'A Compatible SHA Series Design Based on FPGA'. In *Proc. of International Conference on Electrical Engineering/Electronics Computer Telecommunications and Information Technology (ECTI-CON)*. pp. 380–384.

[9] Lin, F. T. and C. Y. Kao. 1995. 'A Genetic Algorithm for Ciphertext-Only Attack in Cryptanalysis'. In *Proc. of International Conference on Systems, Man and Cybernetics*. Vancouver, BC: pp. 650–654.

[10] Abd-Elmonim, W. G., N. I. Ghali, A. E. Hassanien, and A. Abraham. 2011. 'Known-Plaintext Attack of DES-16 Using Particle Swarm Optimization'. In *Proc. of World Congress on Nature and Biologically Inspired Computing (NaBIC)*. pp. 12–16.

[11] Yan, S. Y. 2012. *Computational Number Theory and Modern Cryptography*. Chichester, UK: John Wiley & Sons, Ltd. doi: 10.1002/9781118188606.part2.

[12] Guha, R. K., Z. Furqan, and S. Muhammad. 2007. 'Discovering Man-in-the-Middle Attacks in Authentication Protocols'. In *Proceedings of Military Communications Conference (MILCOM)*. pp. 1–7.

[13] Gautam, T. and A. Jain. 2015. 'Analysis of Brute Force Attack Using TG Dataset'. In *Proc. of SAI Intelligent Systems Conference (IntelliSys)*. pp. 984–988.

[14] Goswami, S., S. Misra, and M. Mukesh. 2014. 'A PKI Based Timestamped Secure Signing Tool for e-Documents'. In *Proc. of International Conference on High Performance Computing and Applications (ICHPCA)*. pp. 1–6.

[15] Hwang, M. S. and L. H. Li. 2000. 'A New Remote User Authentication Scheme Using Smart Cards'. *IEEE Transactions on Consumer Electronics* 46 (1): 28–30.

[16] Hwang, M. S., C. C. Lee, and Y. L. Tang. 2002. 'A Simple Remote User Authentication Scheme'. *Mathematical and Computer Modelling* (Elsevier) 36 (1–2): 103–107.

[17] Das, M. L., A. Saxena, and V. P. Gulati. 2004. 'A Dynamic ID-based Remote User Authentication Scheme'. *IEEE Transactions on Consumer Electronics* 50 (2): 629–631.

[18] Liao, I. E., C. C. Lee, and M. S. Hwang. 2006. 'A Password Authentication Scheme over Insecure Networks'. *Journal of Computer and System Sciences* (Elsevier) 72 (4): 727–740.

[19] Luca, A. D., A. Hang, F. Brudy, C. Lindner, and H. Hussmann. 2012. 'Touch Me Once and I Know It's You!: Implicit Authentication Based on Touch Screen Patterns'. In *Proc. of the ACM SIGCHI Conference on Human Factors in Computing Systems*. pp. 987–996.

[20] Sae-Bae, N., K. Ahmed, K. Isbister, and N. Memon. 2012. 'Biometric-Rich Gestures: A Novel Approach to Authentication on Multi-Touch Devices'. In *Proc. of the SIGCHI Conference on Human Factors in Computing Systems*. pp. 977–986.

[21] Perrig, A., R. Canetti, J. D. Tygar, and D. Song. 2000. 'Efficient Authentication and Signing of Multicast Streams Over Lossy Channels'. In *Proc. of the IEEE Symposium on Security and Privacy*. pp. 56–73.

[22] Perrig, A., R. Canetti, D. Song, and J. Tygar. 2001. 'Efficient and Secure Source Authentication for Multicast'. In *Proc. of Network and Distributed System Security Symposium*.

[23] Refaei, T., M. Horvath, M. Schumaker, and C. Hager. 2015. 'Data Authentication for NDN Using Hash Chains'. In *Proc. of the IEEE Symposium on Computers and Communication (ISCC)*. pp. 982–987.

[24] Agren, M., M. Hell, and T. Johansson. 2012. 'On Hardware-Oriented Message Authentication'. *IET Information Security* 6 (4): 329–336.

[25] Kumar, M., A. Avasthi, and G. Mishra. 2011. 'Advancing the Cryptographic Hash-Based Message Authentication Code'. *IACSIT International Journal of Engineering and Technology* 3 (3): 269–273.

[26] Lampson, B., M. Abadi, M. Burrows, and E. Wobber. 1992. 'Authentication in Distributed Systems: Theory and Practice'. *ACM Transactions on Computer Systems* 10 (4): 265–310.

[27] Gray, R. M. 2006. 'Toeplitz and Circulant Matrices: A Review'. *Foundations and Trends in Communications and Information Theory* 2 (3): 155–239.

[28] Wang, J., X. Shen, and Y. Qi. 2008. 'Method to Implement Hash-Linking Based Content Integrity Service'. In *Proc. of Internal Conference on Genetic and Evolutionary Computing (WGEC)*. pp. 24–27.

[29] Yin, X. C., Z. G. Liu, and H. J. Lee. 2014. 'An Efficient and Secured Data Storage Scheme in Cloud Computing Using ECC-Based PKI'. In *Proc. of International Conference on Advanced Communication Technology, Pyeongchang*. pp. 523–527.

[30] Li, J., J. Li, X. Chen, C. Jia, and W. Lou. 2015. 'Identity-Based Encryption with Outsourced Revocation in Cloud Computing'. *IEEE Transactions on Computers* 64 (2): 425–437.

[31] Tseng, Y. M., T. T. Tsai, S. S. Huang, and C. P. Huang. 2016. 'Identity-Based Encryption with Cloud Revocation Authority and Its Applications'. *IEEE Transactions on Cloud Computing* 99.

[32] Wang, H., Q. Wu, B. Qin, and J. Domingo-Ferrer. 2014. 'Identity-Based Remote Data Possession Checking in Public Clouds'. *IET Information Security* 8 (2): 114–121.

[33] Wang, B., S. S. M. Chow, M. Li, and H. Li. 2013. 'Storing Shared Data on the Cloud via Security-Mediator'. In *Proc. of the IEEE International Conference on Distributed Computing Systems (ICDCS)*. pp. 124–133.

[34] Wang, C., Q. Wang, K. Ren, and W. Lou. 2010. 'Privacy-Preserving Public Auditing for Data Storage Security in Cloud Computing'. In *Proc. of the IEEE INFOCOM*. pp. 1–9.

[35] Zhou, J. and D. Gollmann. 1996. 'A Fair Non-repudiation Protocol'. In *Proc. of the IEEE Conference on Security and Privacy*. pp. 55–61.

[36] Muntean, C., R. Dojen, and T. Coffey. 2009. 'Establishing and Preventing a New Replay Attack on a Non-Repudiation Protocol'. In *Proc. of the IEEE International Conference on Intelligent Computer Communication and Processing (ICCP)*. pp. 283–290.

[37] Zhou, J. and D. Gollmann. 1997. 'An Efficient Non-repudiation Protocol'. In *Proc. of the IEEE Workshop on Computer Security Foundations*. pp. 126–132.

[38] Kremer, S., and J. F. Raskin. 2003. 'A Game-Based Verification of Non-Repudiation and Fair Exchange Protocols'. *Journal of Computer Security* 11 (3): 399–429.

[39] Carroll, C. 2011. 'System and Method for Non-Repudiation within A Public Key Infrastructure'. U.S. Patent US 8082 446 B1, Dec. 20, 2011, US Patent 8,082,446.

[40] Chen, Y. C., G. Horng, and C. L. Liu. 2013. 'Strong Non-Repudiation Based on Certificate-less Short Signatures'. *IET Information Security* 7 (3): 253–263.

[41] Ye, F., Y. Qian, and R. Q. Hu. 2015. 'An Identity-Based Security Scheme for a Big Data Driven Cloud Computing Framework in Smart Grid'. In *Proc. of the IEEE GLOBECOM.* pp. 1–6.

[42] Wen, M., R. Lu, K. Zhang, J. Lei, X. Liang, and X. Shen. 2013. 'PaRQ: A Privacy-Preserving Range Query Scheme Over Encrypted Metering Data for Smart Grid'. *IEEE Transactions on Emerging Topics in Computing* 1 (1): 178–191.

# Part III
Smart Grid Data Management and Applications

CHAPTER 11

# Smart Meter Data Management

It is evident that smart meters will play a key role in managing real-time energy supply–demand and billing customers, as discussed in Chapter 9. Typically, smart meters are expected to be deployed at the distribution side to monitor real-time power usage by the customers. The smart meters' information (such as energy demand) is sent to the service providers to take adequate decisions (such as real-time price of energy and switching on more generators). Consequently, it is required to have adequate mechanisms in place to store, manage, and process the smart meter data. In this chapter, we will learn about different methodologies, which are useful to manage smart meter data and take appropriate decisions in order to establish an improved smart grid environment.

## 11.1 Smart Metering Architecture

Figure 11.1 presents a schematic view of the smart metering architecture, while focusing on different layers in smart meter information collection. Smart meters are deployed at the customers' end to monitor in-house or in-building energy consumption. The data from the smart meters are reported to the utility provider both in real-time and on a day-ahead basis. The information sent by the smart meters are aggregated by multiple aggregators deployed on the distribution side. The collected information is aggregated depending on the content of the information. Further, the aggregated information is stored in the local servers, which are maintained by specific service providers. For simplicity, we can assume that such local servers are maintained by the owners of the micro-grids, in order to have energy consumption information within that micro-grid area. Finally, the information collected from all local servers are stored at the central data center. The data center helps all authenticated users to have real-time energy supply–demand information of all generators and consumers present in the smart grid architecture.

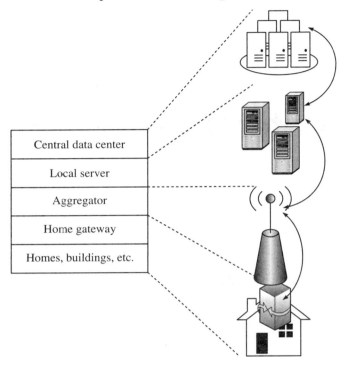

**Figure 11.1**  Schematic view of the smart meter information collection with different layers

## 11.2  Challenges and Opportunities

In this section, we will discuss some of the challenges and opportunities present in managing smart meter data in the smart grid.

### 11.2.1  Requirement of scalable computing facility

In the smart grid, there are at least three different energy periods – on-peak, off-peak, and mid-peak, as discussed in Chapter 5. Therefore, the number of customers in different time periods will vary depending on their energy requirements. Consequently, the number of smart meters reporting the real-time energy consumption information will also vary. This opens up scalability issues in the smart grid, i.e., the system should support varied number of participants in the energy trading process. For example, during on-peak hours, more number of smart meters may participate in the energy trading process. At that time period, the smart grid should be capable of managing information from all the smart meters. Otherwise, in the worst case scenario, the grid may fail to provide electricity services.

In contrast, during the off-peak periods, less number of smart meters will participate in the energy trading process. Therefore, the resource required to manage the information collected from the smart meters is less. Consequently, we need to have a scalable smart meter data management system in order to provide uninterrupted energy service to the customers, while considering the utility of the service providers.

To cope with the issues related to scalability, the integration of cloud computing services with smart grid is considered to be essential. Cloud computing can fulfill multiple requirements of smart grid management. Using the concept of virtualization, resources can be procured and relieved, depending on real-time requirements. For example, as discussed earlier, more number of resources can be procured from third party service providers during on-peak periods to manage the huge information received from the smart meters. On the other hand, these resources can be relieved during off-peak periods, as there are less number of smart meters involved in the energy trading process.

### 11.2.2 Presence of heterogeneous data

In a smart grid, multiple parties participate in the energy trading process. Therefore, the data generated from multiple parties (other than smart meters) should be treated in a different manner. However, in the smart grid architecture, all information would be collected through the same network. Therefore, we have heterogeneous data carried over the same network. Consequently, appropriate data mining policies are also required to handle such heterogeneous data present in the smart grid.

Big data analytics is a useful approach to manage such heterogeneous data. Using different data management tools, smart meter data can be distinguished differently, and appropriate policies can be implemented.

### 11.2.3 Requirement of large storage devices

In addition to scalable computing, the huge requirement of storage devices is an important challenge in the smart grid. The information collected from smart meters are required to be stored in a precise manner before processing them to take appropriate decisions. Consequently, to store the heterogeneous data coming from multiple entities in the smart grid, suitable data analytics tool are required. Further, the distributed nature of the smart grid requires both appropriate distribution of storage devices at all data centers and the local servers as well. Although processing large volume of storage devices is not challenging in the modern era of computer systems, the maintenance of such devices is an important challenge.

### 11.2.4 Information integration from different levels

The integration of information from different levels should be improved, so that information from one department is easily accessible to other departments, considering the associated security parameters. For example, autonomous operation of micro-grids often leads to 'islanded information' [1]. Therefore, a micro-grid generally does not know the status of other micro-grids. As a result, although there is an energy surplus/deficit, micro-grids do not exchange energy among themselves to balance real-time energy supply–demand.

This situation can be avoided by integrating the information from multiple micro-grids on a common platform. For example, a micro-grid that has excess energy supply can exchange real-time energy supply–demand information with another micro-grid that has energy deficit. This can be done through the integration of information from all parties and the processing of them in real-time.

### 11.2.5 Complex architecture in the presence of multiple parties

The integration of the smart grid architecture with the traditional power grid may lead to a highly complex networking system for information management. For example, involving traditional utility providers into the smart grid system for information management may lead to higher cost and inadequate networking policies. Hence, we need to have some specific policies/solution approaches to support the traditional power systems, while integrating bi-directional communication on top of the former.

Managing information and network by separate service providers is a promising approach. For example, the information management and the support of bi-directional network are outsourced from other service providers. In such a scenario, the network and information will be managed by the external entity, so that the traditional utility providers need to focus only on balancing real-time energy supply and demand based on the processed information.

## 11.3 Smart Meter Data Management

In this section, the reader will learn about some of the existing approaches that are useful for smart meter data management in the smart grid.

### 11.3.1 Cluster-based management

Load-profiling is a useful approach to get the day-ahead energy consumption pattern by the customers in the smart grid. For example, the day-ahead energy consumption profile, i.e., the load profile, helps the utility provider to get energy requirements by the customers in different time periods. Consequently, the utility providers are able to take adequate measures to meet the energy consumption requirements by the customers in real-time. To take adequate decisions, clustering of load profiles is one of the promising approaches [2]. Clustering can be done in two ways – direct and indirect. In direct clustering, the customers are grouped according to their energy requirements. For example, customers having high energy demand are present in cluster 1, whereas customers with low energy demand are present in cluster 2. Accordingly, the grid supplies energy to the customers present in both the clusters. This type of clustering method can be used to prioritize customers. For example, the grid can consider customers with high energy demand to be of higher priority, and vice-versa. On the other hand, when the clustering methods are indirectly applied to cluster the load profile, the method is called indirect clustering. For example, clustering can be done according to the energy price. For home customers, the price of energy is different from that of businesses. Consequently, businesses can be given higher priority compared to home customers. Several clustering methods are available such as $k$-means [3], fuzzy $k$-means [4], hierachical clustering [5], self-organizing maps [6], and support vector machines [7], which are useful for clustering the energy consumption profile in smart grid. The performance of each clustering method can be evaluated using the existing methods such as clustering dispersion indicator (CDI), scatter index (SI), and mean index adequacy (MIA) [2, 8].

In the smart grid, the energy consumption pattern varies a lot from one customer to another. Moreover, the energy consumption profile also differs from one day to another for the same customer. Figure 11.2 shows the energy consumption profile for two customers. The large dimension of data collected by smart meters in the smart grid poses challenges to researchers for data storage, processing, and taking appropriate decisions in real-time. Indirect clustering methods are required to be applied to cluster such smart meter data having large dimensions. For example, dimension reduction methods should be applied for reducing the data size before applying any clustering method. Such clustering methods are implemented in two ways – (a) feature extraction-based clustering, and (b) time series-based clustering. The most popular feature extraction-based clustering method is the principal component analysis (PCA) [9, 10]. On the other hand, several methods are available to reduce the dimension of the smart grid load profile, such as the discrete Fourier transform (DFT) [11, 12], symbolic aggregate approximation (SAX) [13], and the hidden Markov model (HMM) [14].

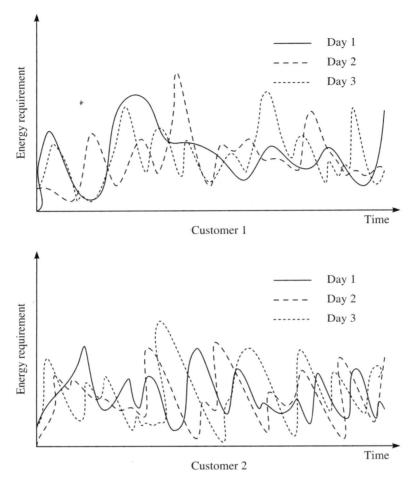

**Figure 11.2** Energy consumption profile for two different customers in smart grid

A Markov decision process-based clustering approach is introduced to formulate the dynamics in the energy consumption profile of customers in smart grid [2]. It is assumed that the energy consumption pattern at a time period always depends on the previous time period, as the behaviors involved in the energy consumption process remains the same for different time periods. The clustering method is executed in four different steps, as follows:

- In the first step, a SAX-based dimension reduction method is applied to reduce the dimensions of the data received from smart meters. Such an approach helps in reducing the storage space and causes ease of communication traffic between smart meters and utility centers, i.e., data centers.
- In the second step, the first search and find density peaks (FSFDP) [15] algorithm is applied to cluster the energy consumption profile received from the customers. The

method reported by Rodriguez and Laio [15] is robust in terms of noise reduction and it has lower time complexity. The dynamics of energy profile for two successive time periods is determined by the K-L distance method [16].

- Third, the FSFDP is integrated with a divide-and-conquer approach to improve the efficiency of the obtained cluster further.
- Finally, based on the results obtained from the clustering scheme, adequate decision is taken to ensure optimal demand response in smart grid.

It is to be noted that the aforementioned scheme can also be applied to cluster a large data set present in big data applications. They are elaborated as follows [2].

- *SAX for load curves*: Using the SAX method, the time series data are converted into symbolic strings. This is done using two steps – piece-wise aggregate approximation (PAA) of load data and the symbolic representation of PAA into discrete strings. Mathematically,

$$x_i = \frac{1}{k_i - k_{i-1}} \sum_{j=k_{i-1}+1}^{k_i} x_j \tag{11.1}$$

and

$$\alpha_i = w_p \text{ if } \beta_{p-1} < x_i < \beta_p \tag{11.2}$$

where the index of the normalized load data is denoted by $j$, the transformed PAA data is denoted by index $i$, the $i$th time domain breakpoint is denoted by $k_i$. Further, $x_i$ is the average value of the $i$th segment. On the other hand, the load curve is symbolically represented as $\alpha$. Finally, $w_p$ is the obtained word from $x_i$. Figure 11.3 presents such a conversion of energy consumption profile into discrete representations.

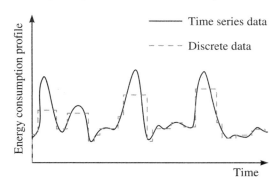

**Figure 11.3** Discrete representation of continuous time series data

- *Time-based Markov model*: As mentioned earlier, the energy consumption pattern in a time period depends on its past time period. To model the dynamic characteristic of the system, discrete Markov model can be applied on symbolic strings obtained from SAX. However, as the customer's load profile varies from day to day, a time-based Markov model is a better option to characterize the dynamic behavior of the load profile of a customer. A transition matrix is formulated, from which a probability transition matrix is obtained. Finally, based on a pre-defined threshold value $\alpha$, the confidence can be measured on the calculated energy consumption profile for the next time period. The test for Markov property is done based on the following equation:

$$\chi^2 = 2 \sum_{i=1}^{N} \sum_{j=1}^{N} f_{ij} |\log \frac{p_{ij}}{p_{*j}}| \tag{11.3}$$

where $p_{*j} = \frac{\sum_{i=1}^{N} f_{ij}}{\sum_{j=1}^{N} \sum_{k=1}^{N} f_{ik}}$, $N$ is the total number of states present in the model, and $f_{ij}$ and $p_{ij}$ are the components of the transition matrix and probability transition matrix, respectively.

- *K-L distance calculation*: The K-L distance between the energy consumption profile of two time periods can be obtained as follows [16]:

$$d_{ij} = \frac{KLD(P_i, P_j) + KLD(P_j, P_i)}{2} \tag{11.4}$$

and

$$KLD(P_i, P_j) = \sum_{n=1}^{s} p_{imn} \log(p_{imn} / p_{jmn}) \tag{11.5}$$

where $P_i$ and $P_j$ are two transition probability matrices with $s$ probability distributions.

- *FSFDP algorithm*: Finally, based on the K-L distance, the profile with minimum distance is selected from all possible distances obtained from Equation (11.4). Therefore, the aforementioned algorithms are useful to cluster the data generated from smart meters, which can in turn be useful to take appropriate decisions.

## 11.3.2 Data compression and pattern extraction

In addition to the smart meter communication and storage of the received data at reduced cost, extraction of useful information from a massive dataset is also a crucial challenge in order to have smooth smart grid operation [17]. Compression is a useful technique to store the smart meters' data. There are two types of compression techniques – lossless and lossy. In lossless compression, the extracted data from the compressed one is identical to the original data. On the other hand, in case of lossy compression, the extracted data is not identical to the original one. However, the extracted information is useful to search important data from a large dataset. The lossy compression technique can be used for load-profiling of the customers. There are several existing schemes, which are useful for feature extraction such as PCA, singular value decomposition (SVD) [18], and DFT. All these feature extraction methods are based on lossy compression.

We will learn about applying $k$-means SVD (K-SVD) algorithm and support vector machine (SVM)-based methods to extract partial usage patterns (PUPs) and to classify the extracted load profiles, respectively.

- K-SVD-based sparse coding: We learned that smart meter data is very sparse, i.e., the load profile of a customer consists of several partial usage patterns (PUPs). Before understanding the actual K-SVD-based algorithm, let us understand the term 'sparse coding'. Sparse coding refers to a signal $x_i = [x_{i,1}, x_{i,2}, \ldots, x_{i,N}]^T$ having $N$ dimensions in a load profile. Further, it can be represented in terms of $K$ basic vectors, which are linear. In smart grids, such vectors are basically considered PUPs. The representation of $x_i$ can be presented as follows:

$$x_i = \sum_{k=1}^{K} a_{i,k} d_k \tag{11.6}$$

or approximated as

$$x_i \approx \sum_{k=1}^{K} a_{i,k} d_k \tag{11.7}$$

where $d_k = [d_{k,1}, D_{k,2}, \ldots, d_{k,N}]^T$ denotes the $k$th PUP, which consists of $N$ dimensions. $a_{i,k}$ denotes the coefficient vector of $K$-PUPs.

For sparse coding for $M$ load profiles, Equation (11.6) can be rewritten in vector form as follows:

$$\mathbf{X} = \mathbf{DA} \tag{11.8}$$

where $\mathbf{X} = [x_1, x_2, \ldots, x_M]$ denotes the set of M load profiles and $\mathbf{A} = [a_1, a_2, \ldots, a_M]$ denotes the coefficient vectors. Therefore, sparse coding can be presented in the form of an optimization problem, as follows:

Minimize $||\mathbf{X} - \mathbf{DA}||_F^2$

subject to

$$||a_i||_0 \leq s_0, \text{ where } 1 \leq i \leq M \tag{11.9}$$

$$a_{i,k} \geq 0, \text{ where } 1 \leq i \leq M \text{ and } 1 \leq k \leq K \tag{11.10}$$

$$d_{k,n} \geq 0, \text{ where } 1 \leq k \leq K \text{ and } 1 \leq n \leq N \tag{11.11}$$

where $s_0$ denotes the maximum number of non-zero elements in each coefficient vector. Equation (11.9) ensures that each load profile follows the target sparcity $s_0$. On the other hand, Equations (11.10) and (11.11) specify that the coefficient vectors and the dictionary are non-negative. Finally, the Frobenius norm, $||*||_F$, in the aforementioned optimization problem is defined as follows:

$$||\mathbf{E}||_F = \sqrt{\sum_i \sum_j |e_{ij}|^2} \tag{11.12}$$

where $e_{ij}$ is the element in $\mathbf{E}$. The Frobenius norm is also called the Euclidean norm, and it is the matrix norm of an $m \times n$ matrix $M$ defined as the square root of the sum of the absolute squares of its elements [19]. Therefore, $\mathbf{E}$ is the square root of the absolute square of its elements.

- Load profile classification: After extracting the PUPs using the feature extraction method explained earlier, the optimal PUP can be selected based on the feature selection and ranking obtained using the SVM-based classification method. In this case, a linear SVM-based classification approach is employed. The feature label pairs $(a_i, y_i)$ are obtained, where $y_i$ corresponds to the label of the $i$th PUP, and is denoted as $[-1, 1]$. The linear SVM optimization problem can be formulated as follows [27]:

$$\underset{\gamma, \omega, \beta}{\text{Minimize}} \frac{1}{2}||\omega||^2 + C \sum_{i=1}^{m} \zeta_i$$

subject to

$$y_i(\omega^T a_i + \beta) \geq 1 - \zeta_i, \; \zeta_i \geq 0 \tag{11.13}$$

where $\omega$ denotes the weights of the features. $C$ is the penalty parameter to take into account the training error, and its value is always positive. $\zeta_i$ denotes the loss function and $\beta$ is the bias factor present in the SVM. Finally, the decision function is defined as follows, while the values of $C$, $\omega$, and $a_i$ are given:

$$f(a_i) = \sin(\omega^T a_i + b) \tag{11.14}$$

Thus, appropriate decisions can be taken using feature extraction and classification methods to predict the load profiles of the customers based on the data from the smart meters.

### 11.3.3 Cloud computing for big data management

**Cloudward: Information management at cloud**

Although cloud technology is expected to offer an optimized framework for smart grid information management, there are several issues that also need to be addressed prior to actual deployment. For example, the distributed architecture of the smart grid requires the use of a distributed optimization framework to deal with heterogeneous and scattered data received from smart meters. Additionally, many service providers (such as network and cloud providers) get involved in the optimization process. Therefore, we need a systematic optimization framework to address such challenges. On the other hand, outsourcing the resources for information management may lead to security and privacy threats to smart grid users. Consequently, adequate security and privacy mechanisms also need to be deployed, so that only authorized users get access to smart grid information. To address such challenges, an optimization framework is introduced based on the cloud computing paradigm, which is useful to reduce the operational cost for smart grid information management [20]. The proposed framework consists of two sub-optimization problems – information storage and computation – that are discussed here.

- Optimization framework for information storage: The optimization problem for information storage using a cloud computing framework can be formulated as follows:

$$\text{Minimize } C = \sum_{d \in D} \sum_{c \in C^D(d)} s^D(d,c) \left( p^S(c) + p^+(d,c) + \sum_{u \in U^D(d)} p^-(c,u) \right)$$

subject to

$$\sum_{c \in C^D(d)} s^D(d,c) = \rho(d) s^D(d), \forall d \in D \tag{11.15}$$

$$s^D(d,c) \leq \frac{\rho(d) s^D(d)}{\theta(d)}, \forall c \in C^D(d), \forall d \in D^S \tag{11.16}$$

$$\frac{s^D(d,c)}{\rho(d) s^D(d)} \in \{0,1\}, \forall c \in C^D(d), \forall d \in D^O \tag{11.17}$$

over $s^D(d,c) \in [0,\infty], \forall c \in C^D(d), \forall d \in D$ \hfill (11.18)

where $C$ is the total cost, $D$ is the set of data items, $D^S$ is the set of data items required to split, $D^O$ is the set of data items required to store in the cloud. $p^+(d,c)$ and $p^-(c,u)$ denote the unit price for uploading data item $d$ to cloud $c$ and the unit price for downloading data from cloud $c$ to user $u$, respectively. Equations (11.15), (11.16), and (11.17) capture the data redundancy constraint, data splitting constraint, and data exclusive constraint, respectively.

- Optimization framework for information computation: Similar to the storage optimization problem, a formulation of an optimization problem for information computation is as follows:

$$\text{Minimize } C = \sum_{d \in D} \sum_{c \in C^D(d)} X^+(d,c) s^D(d) \left( p^S(c) + p^+(d,c) \right)$$

$$+ \sum_{t \in T} \sum_{c \in C^T(t)} X^T(t,c) \left( k(t,c) + \sum_{u \in U^T(t)} S^T(t) + p^-(c,u) \right)$$

$$+ \sum_{t-1 \in T} \sum_{t_2 \in T^\tau(t_1)} \sum_{c_1 \in C^T(t_1)} \sum_{c_2 \in C^T(t_2)} X^I(t_1,t_2,c_1,c_2) S^T(t_1) p^I(c_1,c_2)$$

subject to

$$\sum_{c \in C^T(t)} X^T(t,c) = 1, \forall t \in T \tag{11.19}$$

$$\frac{\sum_{t \in T^D(d)} X^T(t,c)}{|T^D(d)|} \leq X^+(d,c) \leq \sum_{t \in T^D(d)} X^T(t,c), \forall c \in C^D(d), \forall d \in D \tag{11.20}$$

$$X^T(t,c) = 0, \forall c \notin C^T(t), t \in T \tag{11.21}$$

$$\frac{1}{2}(X^T(t_1,c_1)) + (X^T(t_2,c_2)) - \frac{1}{2}X^I(t_1,t_2,c_1,c_2)$$

$$\leq \frac{1}{3}(X^T(t_1,c_1)) + (X^T(t_2,c_2)) + \frac{1}{3}, \forall c_1 \in C^T(t_1) \tag{11.22}$$

$$\forall c_2 \in C^T(t_2), \forall t_2 \in T^\tau(t_1), \forall t_1 \in T \tag{11.23}$$

$$\text{over } X^+(d,c) \in \{0,1\}, \forall c \in C^D(d), \forall d \in D \tag{11.24}$$

$$X^T(t,c) \in \{0,1\}, \forall t \in T, \forall c \in C \tag{11.25}$$

$$X^I(t_1,t_2,c_1,c_2) \in \{0,1\}, \forall c_1 \in C^T(t_1), \forall c_2 \in C^T(t_2), \forall t_2 \in T^\tau(t_1), \forall t_1 \in T \tag{11.26}$$

where Equation (11.19) captures the task execution constraint. Equations (11.20) and (11.21) capture the data upload constraint. On the other hand, Equations (11.22)–(11.25) ensure the inter-cloud intermediate data transfer constraint.

To solve the aforementioned optimization problem, a joint optimization problem can be formulated and solved using mixed integer linear programming. Thus, the optimization problem helps to optimize the cost for information storage and computation on the cloud platform for smart grid data management.

**Smart-Frame: Information management in the cloud**

In addition to scalable information management, security is also an important aspect to consider while managing smart grid information in the cloud. Baek et al. [21] proposed a secure cloud computing framework for smart grid information management, named 'smart-frame'. The proposed framework consists of three hierarchical levels – top, regional, and end-user. The top and regional levels are part of the cloud computing framework; the end-user level corresponds to end-users' smart phones. The proposed scheme utilizes identity-based encryption (IBE) and signature [22, 23], and identity-based proxy re-encryption [24]. Figures 11.4 and 11.5 present schematic views of identity based encryption and signature scheme, respectively.

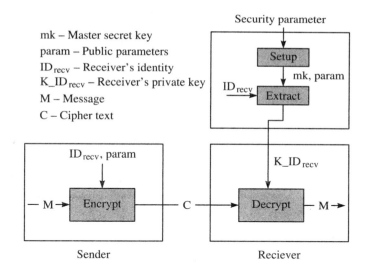

**Figure 11.4** Schematic view of the identity-based cryptosystem

As mentioned earlier, identity-based encryption with signature consists of three levels. Figure 11.6 shows the levels and communication paradigm. Therefore, the end-users exchange information with the regional cloud and the top cloud using the identity-based encryption with signature method scheme, as presented in Figures 11.4 and 11.5. Thus, we have a secure information management framework in the smart grid system.

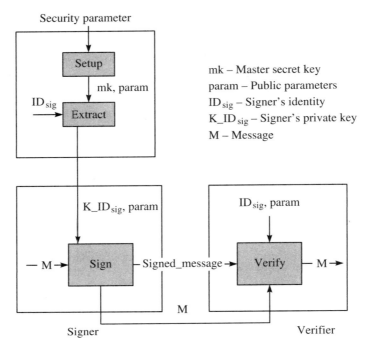

**Figure 11.5** Schematic view of an identity-based signature

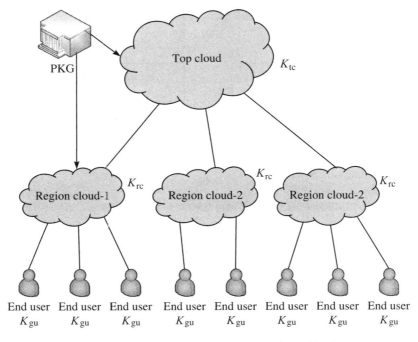

**Figure 11.6** Hierarchical levels of identity-based encryption with signature

## 11.3.4 Calibration with big data

The estimation of electricity distribution parameters can improve smart grid energy management. Smart meters send energy consumption information to the utility providers; this information is stored in data centers. We discussed earlier that the estimation of load profile for customers helps utility providers to improve energy reliability with reduced cost. From this perspective, a big data-based estimation model was proposed by Peppanen et al. [25] to estimate the electricity distribution system parameters for improved energy management in the smart grid, as shown in Figure 11.7.

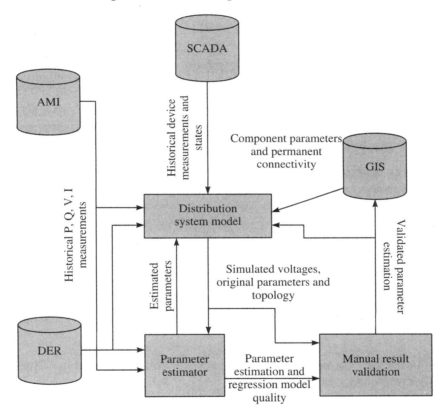

**Figure 11.7** Big data for distribution parameter estimation framework

As presented in Figure 11.7, AMI provides the load profiles; GIS provides the current model components, parameters and permanent connectivity; DER provides distributed generation profiles. The system is expected to be automated with very limited human intervention for estimating the system parameters. Finally, the parameter estimation algorithm is evaluated, as presented in Figure 11.8 (adopted from [25]). Thus, using the big data approach, the distribution system parameters can be estimated for improved decision making in the smart grid system.

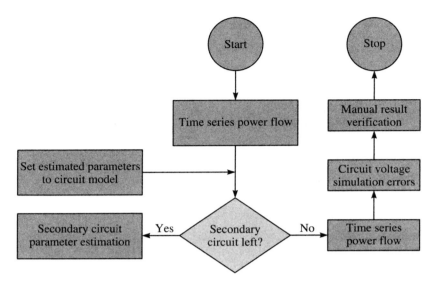

**Figure 11.8** Algorithm for parameter estimation

### 11.3.5 Fuzzy logic-based management

Outage management in the electric distribution system is another important aspect of the smart grid. Weather has great impact on power outage in the electric distribution system [26]. Therefore, outage prediction based on predictive risk analysis of weather forecast can be useful for improved decision making in the smart grid. However, traditional outage management schemes suffer from several limitations:

- In the traditional outage management system, mostly customers call the utility provider to inform about any outage occurring at specific locations. Therefore, the information may be incomplete and non-trustworthy.
- Due to the absence of accurate distribution network models, it is difficult to know about individual customers' outages. As a result, priority of customers' requirements may not be considered in a fair manner.
- Manual prediction models often lead to inaccurate outage management policies, which, in turn, increases the duplicity in the restoration work.
- Finally, asset management is not good due to the use of aged equipment, which causes higher failure rate in peak hours.

A fuzzy logic-based predictive risk analysis framework is available for distribution outage management [26]. This framework consists of three phases – predictive risk analysis, real-time operations, and post-event operations. Figure 11.9 presents this model for predictive risk analysis in the smart grid. The presented model can be used to predict outage in electric distribution systems, and adequate measures can be taken to improve the reliability of energy service to the customers in the smart grid.

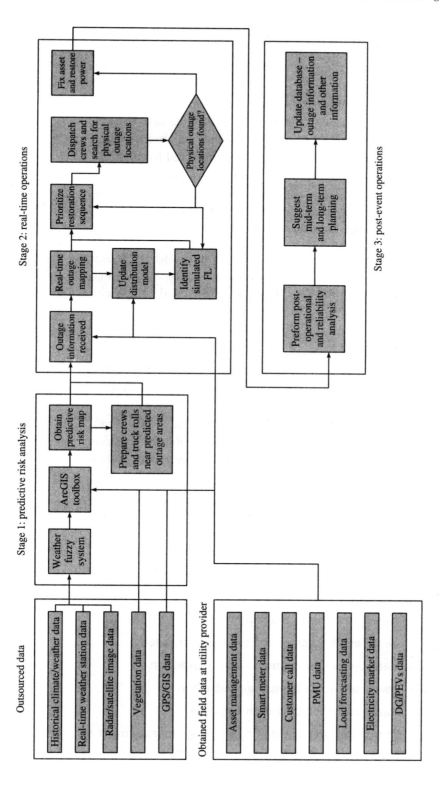

**Figure 11.9** Framework for outage management in the smart grid (adopted from [26])

## 11.4 Future Trends and Issues

Some of the future trends and issues are identified based on the existing schemes mentioned earlier.

- We see that environmental parameters also influence the electrical distribution system. Therefore, different environmental parameters such as temperature, humidity, and weather forecasting can be considered for predicting the load profiles for customers, in addition to traditional parameters. For example, for outage management in smart grid, weather conditions may be considered. This is also true for the load profile prediction model. This factor can be considered for improved decision making in the smart grid.

- The sparse variety is considered to be uniform for all cases, as presented in Section 11.3.2. However, different values for sparse variety may be considered to get improved results, while considering the associated complexity in the process.

## 11.5 Summary

In this chapter, different schemes for smart meter data management were discussed from the perspective of big data and data management. Smart meter data management was presented, while focusing on the challenges from big data perspectives. Different schemes, which are useful for smart meter data management and to take improved decisions in the smart grid systems, were presented. Finally, few future research trends and issues were discussed based on the limitations of the existing schemes.

### Test Your Understanding

Q01. Describe the smart metering architecture.
Q02. Describe the challenges and opportunities in managing smart meter data in the smart grid.
Q03. Explain cluster-based data management.
Q04. Explain the different steps in the clustering methods.
Q05. Describe data compression and pattern extraction.
Q06. Explain how information is managed at cloud.
Q07. What is meant by calibration with big data?
Q08. Mention the limitations of traditional outage management scheme.

Q09. State the three phases of fuzzy logic-based predictive risk analysis framework.

Q10. Mention some future trends and issues of the existing schemes of smart meter data management.

# References

[1] Fang, X., S. Misra, G. Xue, and D. Yang. 2012. 'Smart Grid – The New and Improved Power Grid: A Survey'. *IEEE Communications of Surveys and Tutorials* 14 (4): 944–980.

[2] Wang, Y., Q. Chen, C. Kang, and Q. Xia. 2016. 'Clustering of Electricity Consumption Behavior Dynamics Toward Big Data Applications'. *IEEE Transactions on Smart Grid* 7 (5): 2437–2447.

[3] Li, R., C. Gu, F. Li, G. Shaddick, and M. Dale. 2015. 'Development of Low Voltage Network Templates Part I: Substation Clustering and Classification'. *IEEE Transactions on Power Systems* 30 (6): 3036–3044.

[4] Zhou, K. L., S. L. Yang, and C. Shen. 2013. 'A Review of Electric Load Classification in Smart Grid Environment'. *Renewable and Sustainable Energy Review* 24: 103–110.

[5] Tsekouras, G. J., P. B. Kotoulas, C. D. Tsirekis, E. N. Dialynas, and N. D. Hatziargyriou. 2008. 'A Pattern Recognition Methodology for Evaluation of Load Profiles and Typical Days of Large Electricity Customers'. *Electric Power Systems Research* 78 (9): 1494–1510.

[6] Verdu, S. V., M. O. Garcia, C. Senabre, A. G. Marin, and F. J. G. Franco. 2006. 'Classification, Filtering, and Identification of Electrical Customer Load Patterns through the Use of Self-Organizing Maps'. *IEEE Transactions on Power Systems* 21 (4): 1672–1682.

[7] Chicco, G., and I. S. Ilie. 2009. 'Support Vector Clustering of Electrical Load Pattern Data'. *IEEE Transactions on Power Systems* 24 (3): 1619–1628.

[8] Chicco, G. 2012. 'Overview and Performance Assessment of the Clustering Methods for Electrical Load Pattern Grouping'. *Energy* 42 (1): 68–80.

[9] Koivisto, M., P. Heine, I. Mellin, and M. Lehtonen. 2013. 'Clustering of Connection Points and Load Modeling in Distribution Systems'. *IEEE Transactions on Power Systems* 28 (2): 1255–1265.

[10] Chicco, G., R. Napoli, and F. Piglione. 2006. 'Comparisons Among Clustering Techniques for Electricity Customer Classification'. *IEEE Transactions Power Systems* 21 (2): 933–940.

[11] Zhong, S. and K. S. Tam. 2015. 'Hierarchical Classification of Load Profiles Based on their Characteristic Attributes in Frequency Domain'. *IEEE Transactions on Power Systems* 30 (5): 2434–2441.

[12] Torriti, J. 2014. 'A Review of Time Use Models of Residential Electricity Demand'. *Renewable and Sustainable Energy Review* 37: 265–272.

[13] Notaristefano, A., G. Chicco, and F. Piglione. 2013. 'Data Size Reduction with Symbolic Aggregate Approximation for Electrical Load Pattern Grouping'. *IET Generation, Transmission and Distribution* 7 (2): 108–117.

[14] Albert, A. and R. Rajagopal. 2013. 'Smart Meter Driven Segmentation: What your Consumption Says about You'. *IEEE Transactions on Power Systems* 28 (4): 4019–4030.

[15] Rodriguez, A., and A. Laio. 2014. 'Clustering by Fast Search and Find of Density Peaks'. *Science* 334 (6191): 1492–1496.

[16] Kullback, S. and R. A. Leibler. 1951. 'On Information and Sufficiency'. *The Annals of Mathematical Statistics* 22 (1): 79–86.

[17] Wang, Y., Q. Chen, C. Kang, Q. Xia, and M. Luo. 2016. 'Sparse and Redundant Representation-Based Smart Meter Data Compression and Pattern Extraction'. *IEEE Transactions on Power Systems*. 32 (3): 2142–2151.

[18] de Souza, J. C. S., T. M. L. Assis, and B. C. Pal. 2017. 'Data Compression in Smart Distribution Systems via Singular Value Decomposition'. *IEEE Transactions on Smart Grid* 8 (1): 275–284.

[19] Golub, G. H. and C. F. Van Loan. 1996. *Matrix Computations*, 3rd edition. Baltimore, MD: Johns Hopkins.

[20] Fang, X., D. Yang, and G. Xue. 2013. 'Evolving Smart Grid Information Management Cloudward: A Cloud Optimization Perspective'. *IEEE Transactions on Smart Grid* 4 (1): 111–119.

[21] Baek, J., Q. H. Vu, J. K. Liu, X. Huang, and Y. Xiang. 2015. 'A Secure Cloud Computing Based Framework for Big Data Information Management of Smart Grid'. *IEEE Transactions on Cloud Computing* 3 (2): 223–244.

[22] Boneh, D. and M. K. Franklin. 2001. 'Identity-Based Encryption from the Weil Pairing'. In *Proc. of Annual International Cryptology Conference*: 213–229.

[23] Shamir, A. 1984. 'Identity-Based Cryptosystems and Signature Schemes. In *Proc. CRYPTO Adv. Cryptol.* 196: 47–53.

[24] Green, M. and G. Ateniese. 2007. 'Identity-Based Proxy Re-encryption'. In *Proc. 5th Int. Conf. Appl. Cryptograph. Netw. Security* 4521: 288–306.

[25] Peppanen, J., M. J. Reno, R. J. Broderick, and S. Grijalva. 2016. 'Distribution System Model Calibration With Big Data from AMI and PV Inverters'. *IEEE Transactions on Smart Grid* 7 (5): 2497–2506.

[26] Chen, P. C. and M. Kezunovic. 2016. 'Fuzzy Logic Approach to Predictive Risk Analysis in Distribution Outage Management'. *IEEE Transactions on Smart Grid* 7 (6): 2827–2836.

[27] Chang, K. W., C. J. Hsieh, and C. J. Lin. 2008. 'Coordinated Descent Method for Large-scale L2-loss Linear Support Vector Machines'. *Journal of Machine Learning Research* 9 (7): 1369–1398.

# CHAPTER 12

# PHEVs: Internet of Vehicles

Plug-in hybrid electric vehicles (PHEVs) are attracting the interest of researchers due to their unique features such as real-time demand side management and green technology. PHEVs use a combination of electric energy and traditional fossil fuel. They have an electric motor and an internal combustion engine. It is expected that PHEVs will produce less amount of air pollution compared to traditional vehicles. Moreover, PHEVs can charge and discharge their batteries, depending on the real-time energy status of the grid, which is widely termed as *vehicle-to-grid* (V2G) and *grid-to-vehicle* (G2V) transfer.

Vehicle-to-grid (V2G) transfer is defined as the energy transfer from an electric vehicle to the power grid. In smart grid, the energy supply from generators and demand from its customers are dynamic over time. Therefore, sometimes, the grid may face supply–demand imbalance, which, in turn, may lead to grid failure. Using the concept of V2G transfer, energy can be supplied to the grid from the vehicle when there is a deficit in energy supply to the grid compared to the energy demand from its customers. Therefore, sudden energy demand peaks can be avoided using such technology.

Vehicles also consume energy to charge their batteries. Consequently, they can charge their batteries when there is a surplus in energy supply to the grid. This energy transfer technology is termed as grid-to-vehicle (G2V) transfer. Therefore, surplus energy can be stored in the PHEVs' batteries without wasting it. Hence, V2G and G2V techniques play an important role in real-time energy management in the smart grid. It is noteworthy that PHEVs are also integrated with different communication technologies, in order to have real-time energy supply–demand information. Consequently, suitable communication technology is also required for efficient operation of the PHEVs, while integrating them with the smart grid.

## 12.1 Convergence of PHEVs and Internet of Vehicles

Internet-of-vehicles (IoVs) are expected to play a major role in real-time information sharing and gathering information about roads and traffic. It is also useful to process, compute, and release the gathered information to other platforms. Based on the received data, the system can guide its users to take appropriate decisions. Moreover, it can provide Internet connectivity to its users as is done in vehicular ad-hoc networks (VANETs).

Today's PHEVs will become tomorrow's IoVs with the energy management facilities. This is because, in addition to energy management, PHEVs can fulfill the objectives of the concept of IoVs. The several advantages of integrating PHEVs with IoVs are as follows:

- PHEVs have low emissions compared to traditional vehicles. Consequently, PHEVs can help in establishing a $CO_2$ reduced environment in the smart grid. Due to the mobile nature of the PHEVs, they can roam around different places. Hence, PHEVs can be used to disseminate information in addition to participating in energy trading in the smart grid.
- Due to efficient fuel utilization, PHEVs are one of the most fuel-efficient cars on the roads when there is traffic.
- PHEVs have bi-directional communication facility to communicate with the utility provider. The same communication technology can be used for general information dissemination such as road condition and traffic condition.
- Finally, PHEVs can also be integrated with smart meters to have home energy management during *parking* hours. Typically, PHEVs are parked at the office premises during office hours and at home during night. Therefore, they can form part of the building energy management system during office hours, and the home energy management system during night.

Figure 12.1 presents the different services that can be obtained from PHEVs. Due to the aforementioned reasons, PHEVs will lead to the establishment of connected vehicles in the near future with the unique feature of energy management. Consequently, efficient network operation is desired as the PHEVs are expected to carry heterogeneous information. To deal with heterogeneous data, big data technologies are applied to analyze the gathered information and to take appropriate decisions. In the next section, several existing schemes that are useful for PHEVs management are discussed.

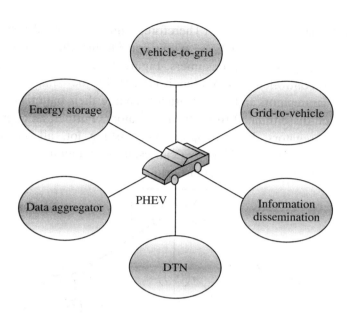

**Figure 12.1**  Different services that can be obtained from a PHEV

## 12.2  Electric Vehicles Management

Recently, several schemes were proposed for electric vehicle management in smart grid from different perspectives – charging and discharging of PHEVs, energy management in smart grid, and on-board Internet service in the presence of PHEVs. These schemes will be discussed in this section.

### 12.2.1  Charging and discharging of PHEVs

The concepts of vehicle to grid (V2G) and grid to vehicle (G2V) transfer have already been discussed in Section 12.1. PHEVs can play an important role in the smart grid by charging/discharging their batteries, while considering the real-time energy supply–demand status at the grid.

#### Optimal scheduling
Due to the inherent features of PHEVs, they can charge and discharge in a coordinated manner in order to build a huge energy storage system in the smart grid. For example, PHEVs can charge their batteries when there is excess energy supply to the grid compared to the total energy demand from customers. On the other hand, PHEVs can discharge their batteries when there is a deficit in the energy supplied from the traditional energy sources, as discussed earlier. Consequently, PHEVs need to know the real-time status of energy supply–demand to the grid. Typically, PHEVs are communicated/controlled by the

aggregator units in the smart grid [1]. Therefore, aggregators need suitable scheduling policy for charging/discharging PHEVs, while considering the energy supply/demand from traditional energy generators/consumers. Figure 12.2 presents a schematic view of such an aggregator-controlled PHEVs environment in the smart grid. PHEVs are associated with an aggregator, and they charge/discharge their batteries according to the real-time information disseminated from the aggregators. Further, an aggregator can also control multiple aggregators. For example, in a parking lot, all the aggregators are controlled by another aggregator. Thus, we can establish a hierarchical and coordinated charging/discharging of PHEVs in the smart grid.

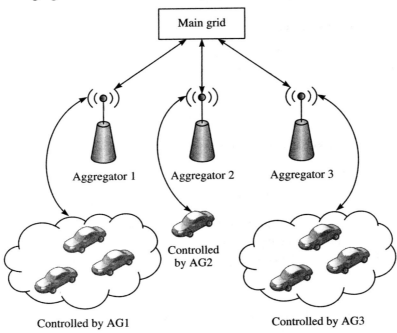

**Figure 12.2** Schematic view of aggregator-controlled PHEVs in a smart grid

To optimally schedule the charging/discharging process of the PHEVs, aggregators need to take optimal decisions based on the real-time energy supply and demand to the grid. A forecasting-based scheduling policy can be employed in the smart grid to charge/discharge the batteries of PHEVs [1]. The charging and discharging power of a PHEV can be represented mathematically as follows:

$$P_{i,\text{dis}} \leq P_i(t) \leq P_{i,\text{char}}, \, t \in [T_{i,\text{in}}, T_{i,\text{out}}] \tag{12.1}$$

where $P_{i,\text{dis}}$ and $P_{i,\text{char}}$ are the maximum discharging and charging capacity of a PHEV, $i$, respectively. $P_i(t)$ is the charged or discharged power of the PHEV at time $t$, which is determined by the IN and OUT time of the PHEV to/from a charging/discharging station in the smart grid. Therefore, we have two scenarios for charging and discharging of PHEVs. Mathematically,

$$P_i(t) = \begin{cases} > 0, & \text{for charging} \\ < 0, & \text{for discharging} \\ = 0, & \text{Otherwise} \end{cases} \quad (12.2)$$

Consequently, the profile of total power of PHEVs can be computed as follows [1]:

$$P_{tot}(\tau) = \mathcal{R}(\tau) + P_A(\tau) \quad (12.3)$$

where $\mathcal{R}(\tau)$ is the total regulated power during the entire participation period $\tau$, and $P_A(\tau)$ is the total aggregated power from PHEVs. Further, the aggregated power $P_A(\tau)$ can be represented as follows:

$$P_A(\tau) = \sum_{i \in N} P_i(\tau) \quad (12.4)$$

Therefore, the objective of the aggregators is to minimize the variance in the total power $P_A(\tau)$ for charging and discharging of the PHEVs [1]. Mathematically,

$$\underset{P_N(\tau)}{\text{Minimize}} \; \mathcal{U}_f(P_A(\tau))$$

subject to

$$P_{i,\text{dis}} \leq P_i(t) \leq P_{i,\text{char}}, \; t \in [T_{i,\text{in}}, T_{i,\text{out}}] \quad (12.5)$$

$$\text{SOC}_i(T_j) \geq \text{SOC}_{i,\text{minCh}} \quad (12.6)$$

$$\text{SOC}_{i,\text{min}} \leq \text{SOC}_i(T_k) \leq \text{SOC}_{i,\text{max}} \quad (12.7)$$

where $\mathcal{U}_f(P_A(\tau))$ denotes the variance in the total charging and discharging power of the PHEVs to the aggregators. Equation (12.5) denotes the charging and discharging limit of a PHEV, while Equation (12.6) shows that the state-of-charge (SOC) of a PHEV should be greater or equal to the minimum SOC, before it can plug-out from a charging station. On the other hand, Equation (12.7) ensures that the SOC is always within the limit of minimum and maximum charge capacity of the PHEV. Further, the variance in the total power can be calculated as follows:

$$\mathcal{U}_f(P_A(\tau)) = \text{Var}(P_{tot}(\tau))$$

$$= \frac{1}{N_T} \sum_{T_k \in \tau} \left( \mathcal{R}_f(T_k) + P_A(T_k) - \frac{1}{N_T} \left( \sum_{T_j \in \tau} (\mathcal{R}_f(T_j) + P_A(T_j)) \right) \right)^2 \quad (12.8)$$

Therefore, based on the forecasted energy profile, the appropriate decision can be taken in order to optimally schedule the charging and discharging of the PHEVs. However, in a real environment, the forecasted information is not always accurate [1]. Therefore, although the aforementioned optimization can produce the best results theoretically, it may not be useful from a practical point of view. Consequently, an online optimization model is useful to take appropriate decisions based on the forecasted information. Mathematically, it can be modeled as an optimization problem as follows [1]:

$$\underset{Q_N(T_k)}{\text{Minimize}}\ \mathcal{U}_0(Q_A(T_k))$$

subject to

$$\eta(P_i(T_k))P_i(T_k) + \eta(FP_i(T_k))FP_i(T_k)$$
$$\geq \frac{C_i}{\Delta t}(\text{SOC}_{i,\text{minCh}} + \text{SOC}_{i,\text{MOS}}(T_k) - \text{SOC}_i(T_k - 1)) \quad (12.9)$$

and Equations 12.5 and 12.7

where $Q_N(T_k)$ denotes the set of all scheduled PHEVs for charging or discharging in $T_k$ time period. $\text{SOC}_{i,\text{MOS}}(T_k)$ denotes the 'margin of safety' of the PHEV. The margin of safety is defined as the maximum energy that can be charged or discharged while considering the battery capacity. $FP_i(T_k)$ denotes future charging and discharging of the $i$th PHEV. Gradient projection method is applied to solve this optimization problem [1]. We see that this approach is useful to deal with the scheduling issues in charging and discharging of PHEVs' batteries in the smart grid, which, in turn, establishes a balanced smart grid environment in the presence of PHEVs.

### Charging and discharging through cooperation

Cooperative charging and discharging of PHEVs' batteries is another useful approach for real-time energy management in the smart grid system. Using this approach, PHEVs and the grid form multiple groups according to their mutual understanding; they can then charge/discharge their batteries in a coordinated manner. Figure 12.3 shows cooperative charging/discharging of PHEVs in a smart grid environment.

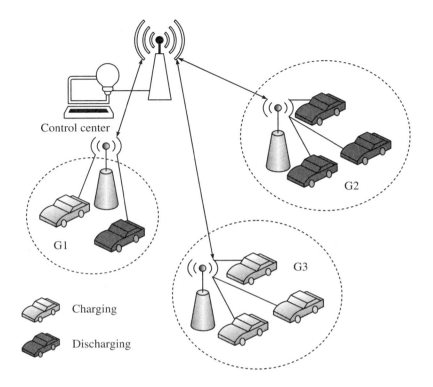

**Figure 12.3** Coordinated charging/discharging of PHEVs while forming different groups

The set of all PHEVs in the smart grid is denoted as $\mathcal{V} = \{V_1, V_2, \ldots, V_n\}$, $n \in \mathcal{N}$. The PHEVs form different coalitions with an aim to maximize their individual payoffs. Mathematically, the set of coalitions can be represented as $\mathcal{G} = \{G_1, G_2, \ldots, G_l\}$, $l \leq n$. Therefore, $|G_i|$ represents the number of PHEVs present in the $i$th coalition. The PHEVs calculate their individual payoff values, based on which, decisions for joining/leaving a coalition are taken. The payoff values are determined by the utility of the PHEVs, which is mathematically determined as [2]:

$$\mathcal{U}_i^t(V_i) = \begin{cases} w_i^t \log(1 + x_i^t) - P_c^t x_i^t, & 0 \leq x_i^t \leq 1 \\ w_i^t (\log(2 + x_i^t) - 1) + P_{d,k}^t x_i^t, & -1 \leq x_i^t < 0 \end{cases} \quad (12.10)$$

where $w_i^t$ denotes the willingness of charging or discharging of the PHEV $i$ at time $t$. The willingness of charging and discharging can be determined based on the state-of-charge (SOC) of the PHEVs' batteries. $x_i^t$ is the normalized electricity to charge or discharge PHEV $i$ at time $t$. $P_c^t$ and $P_{d,k}^t$ denote the charging and discharging price at time $t$, respectively. It is to be noted that the charging price at a particular time is kept the same for all PHEVs. On the other hand, the discharging price at a particular time can be different for different PHEVs. Therefore, we have the component $k$ in the discharging price $P_{d,k}^t$. The objective is to choose an adequate value for $x_i^t$ for which the utility is maximized, while considering $w_i^t$, $P_c^t$, and $P_{d,k}^t$. According to Yu et al. [2], different actions (presented in Figure 12.4) can be taken to charge or discharge PHEVs. A coalition game-theoretic approach may be adopted to solve this problem. In this approach, multiple groups are formed among the PHEVs, and they take decisions in a cooperative manner. The coalition game-theoretic approach can be segmented into three stages – initialization, coalition formation, and decision making – which are briefly explained here.

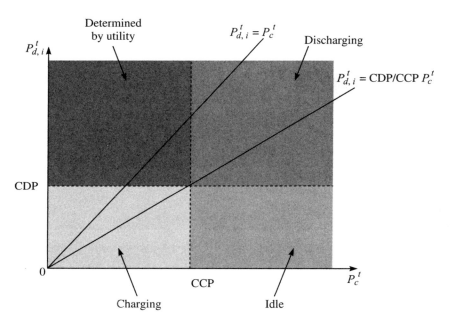

**Figure 12.4** Different actions to charge/discharge the PHEVs

- *Initialization*: In the initialization phase, all the PHEVs act individually; they receive charging and discharging price information from the grid. After receiving the price information, the PHEVs advertise themselves either as *inviters* or *invitees*. The term *inviters* is used for the PHEVs that are interested in charging their batteries. On the other hand, the term *invitees* is used for PHEVs that are interested in discharging their batteries. *Inviters* are interested in forming coalition with the *invitees* so that the energy supply and demand from the PHEVs are balanced. Therefore, *inviters* starts the coalition forming process. When invitees accept the request from *inviters*, a coalition is formed between them.
- *Coalition formation*: In the coalition formation process, the PHEVs calculate their utility values to match with optimal peers of *inviters–invitees*, so that their individual payoff is maximized. The coalition formation process is further subdivided into three steps – *update preference list*, *match inviter–invitee*, and

*merge and split*. In the *update preference list*, the PHEVs advertise their individual preference list to cooperate with other PHEVs. Based on the *preference list*, the *inviter–invitee* matching is done. Finally, *merge and split* operations are executed based on the optimal value of the utility, i.e., merging and splitting of coalitions are done in an iterative manner until a stable condition is reached. In the stable condition, none of the PHEVs can increase their payoff values by changing their coalitions. This stable condition is known as *equilibrium* condition in game theory.

- *Decision making*: In the decision making process, following situations may arise: (1) total amount of charged energy and discharged energy is the same, so that no energy exchange is required outside of the coalition; (2) total charged energy is higher than the total discharged energy. Therefore, the deficit amount of energy is required to be purchased from the main grid; (3) total discharged energy is higher than the total charged energy, which results in excess energy to be sold back to the grid.

Through such processes, we can execute coordinated charging and discharging of PHEVs in the smart grid.

**Resource reservation**

Resource reservation has been a useful technique in networking systems for several years. In the smart grid environment, typically, micro-grids are responsible for distributing electricity to the customers in a combination of both renewable and non-renewable energy sources. Therefore, a micro-grid can be treated as a small substation that has its own energy supply and demand patterns from the generators and customers, respectively. Consequently, day-ahead energy supply–demand curve can be obtained easily, which, in turn, reveals the energy excess/deficit in different time periods throughout the day. PHEVs are mobile in nature, and also charge/discharge their batteries at different time periods in a day, depending on their requirements. Therefore, PHEVs can reserve the energy resources of a micro-grid to charge batteries in a particular time period in advance considering the energy curve in that time period. They can also reserve a time period, in which they are willing to discharge their batteries considering the energy curve of the micro-grid. Therefore, we can have well-balanced charging and discharging scenarios in a micro-grid using the reservation process. Towards this objective, Kaur et al. [3] proposed a novel resource reservation scheme in the smart grid for charging and discharging the batteries of PHEVs. Figure 12.5 depicts a schematic view of a resource reservation-based scheme in the smart grid in the presence of PHEVs (adopted from [3]).

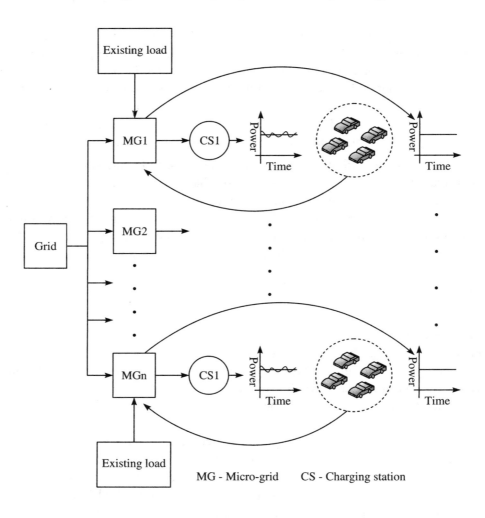

**Figure 12.5** Schematic view of resource reservation in a smart grid in the presence of PHEVs

As depicted in Figure 12.5, the objective of the micro-grid is to balance the energy supply–demand by regulating the PHEVs in different time periods, in order to minimize the on-peak and off-peak hours [3]. Mathematically,

$$\text{Minimize} \sum_{i=1}^{n} C_i^d S_i^d$$

subject to

$$\sum_{i=1}^{n} S_i^d(E_r) = E_{\text{req}} \tag{12.11}$$

where $C_i^d$ is the discharging cost offered by the PHEV $i \in \mathcal{N}$. $S_i^d$ is the state-of-charge which can be discharged by the PHEV $i$. $E_r$ and $E_{\text{req}}$ are the rated capacity of the PHEV and the required energy by the micro-grid, respectively. It is to be noted that the optimization problem can give optimized solution for peak-saving, which is similar to the knapsack problem [4].

## 12.2.2 Energy management for data centers and PHEVs

Due to the advances in bi-directional communication network technology, it is expected that there will be multiple data centers to manage real-time data sent by multiple parties (such as smart meters) in the smart grid. Therefore, the data centers play an important role in information management, which, in turn, helps in the decision making process in real-time. However, a huge amount of energy is also required to run such smart grid data centers. Moreover, data centers are to be regulated according to specific requirements. PHEVs can play an important role in regulating the energy service at the data centers, while considering different geographical locations of the data centers and mobile nature of the PHEVs. Figure 12.6 presents the system model in which energy is regulated in the presence of data centers and PHEVs (adopted from [5]).

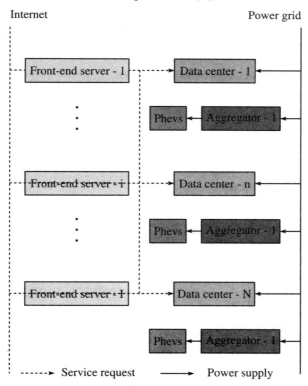

**Figure 12.6** Schematic system model of energy regulation with data centers and PHEVs

As depicted in Figure 12.6, the aggregator schedules the charging process of PHEVs, so that the data centers are not affected by the presence of PHEVs in a particular geographical region. Additionally, the aggregators and data centers are fed with energy supply from the power grid to conduct normal operations. In this model, different sub-models are used, such as the model associated with the data centers and the front-end servers, and the model associated with PHEVs. The mathematical model associated with data centers and front-end servers may be as follows [5]:

$$\sum_{n=1}^{N} E_{i,n,t} = \lambda_{i,t}, \forall i,t \tag{12.12}$$

and

$$E_{i,n,t} \geq 0, \forall i,n,t \tag{12.13}$$

where $E_{i,n,t}$ is the interactive workload distributed between the $i$th front-end server and the $n$th data center at time $t$. Therefore, $E_{i,n,t}$ is always greater than or equal to zero. $\lambda_{i,t}$ is the total amount of interactive workload coming from front-end servers, which are distributed among all available data centers. Besides the workload from front-end servers, multiple back-end workloads are also present, which are required to be considered in the system model. Mathematically, the back-end workload should abide by the following inequality:

$$0 \leq \pi_{n,t} \leq \hat{\pi}_{n,t}, \forall n,t \tag{12.14}$$

where $\hat{\pi}_{n,t}$ is the maximum workload in the back-end server and $\pi_{n,t}$ is the distributed workload.

Finally, the power consumption at the data center can be calculated as follows:

$$P_{n,t}^{d} = \alpha_n + \beta_n \left( \sum_{i=1}^{I} E_{i,n,t} + \pi_{n,t} \right), \forall n,t \tag{12.15}$$

where $\alpha_n = C_n(P_n^{\text{idle}} + (P_{\text{eff},n} - 1)P_n^{\text{peak}})$, and $\beta_n = P_n^{\text{peak}} - P_n^{\text{idle}}$. Further, $P_{\text{eff},n}$ denotes the effective power usage by the $n$th data center. $P_n^{\text{idle}}$ and $P_n^{\text{peak}}$ denote the power requirement during the idle and peak conditions, respectively.

We now discuss the models associated with PHEVs. PHEVs send their charging requests to the aggregators, which are to be served based on a first-in-first-out (FIFO) policy. Based on the FIFO logic, the queue for charging requests from PHEVs is modeled as follows:

$$Q_{n,q,t+1} = \max[Q_{n,q,t} - x_{n,q,t}, 0] + a_{n,q,t}, \forall n,q,t \tag{12.16}$$

where $x_{n,q,t}$ and $a_{n,q,t+1}$ are the served and arrived requests in queue at time $t$, respectively. Further, the requested energy demand should be within a specified limit, otherwise, it cannot be served. This is represented mathematically as follows:

$$0 \leq x_{n,q,t} \leq \min\{x_{n,q}^{\max}, Q_{n,q,t}\}, \forall n, q, t \tag{12.17}$$

As mentioned earlier, if the energy demand is above the specified value, it can be dropped by the data center. Therefore, the total served energy demand is always within the peak limit. Mathematically,

$$\sum_{q=1}^{Q_n} (x_{n,q,t} - u_{n,q,t}) + P_{n,t}^d \leq P_n^{\text{limit}}, \forall n, t \tag{12.18}$$

where $u_{n,q,t}$ is the dropped energy demand by the data center $n$ at time $t$. Further, the energy demand in the queue should be served within the limited time period.

Finally, considering all the constraints presented in Equations (12.12) through (12.18), the optimization problem for cost minimization is modeled as follows:

$$\text{Minimize} \limsup_{T \to \infty} \frac{1}{T} \sum_{t=0}^{T-1} \mathbb{E}\{\tau_t\}$$

subject to

Equations 12.12 to 12.18

$$0 \leq u_{n,q,t} \leq x_{n,q,t} \tag{12.19}$$

where

$$\tau_t = \sum_{n=1}^{N} \left( S_{n,t} \left( P_{n,t}^d + \sum_{q=1}^{Q_n} (x_{n,q,t} - u_{n,q,t}) \right) \right) \tag{12.20}$$

and $S_{n,t}$ is the electricity price associated with data center $n$ at time $t$. The aforementioned optimization problem can be solved using the Lyapunov optimization technique [6] in addition to the alternating direction method of multipliers (ADMM) method [7].

### 12.2.3 Providing on-board internet service facility

Due to the inherent features of PHEVs (such as green energy, V2G and G2V facilities), the number of PHEVs is increasing regularly. Therefore, PHEVs can also build a network (similar to VANETs) through which they can communicate with one another, and take coordinated decisions. Moreover, at present, vehicles are facilitated with on-board Internet service through which the drivers can get adequate information about road conditions, which, in turn, maximizes the safety and comfort of passengers and minimizes the travel time to reach a destination [8]. However, the vehicles should be capable of handling adverse situations (such as heavy rain and earthquake) in a suitable manner. Moreover, it must be secure from unauthorized access and attackers. Let us discuss some emerging applications of vehicles facilitated with Internet of service [8].

- *Content validity*: The vehicles generate different information (termed as content) that are used for decision making by the vehicle itself. Further, the generated content is disseminated among other interested vehicles so that all vehicles get benefited. Typically, the content has a time–space validity, within which it is useful. For example, a speed breaker content is valid only for vehicles that approach a bad road. Therefore, restriction can be made based on the distance from the road. For example, vehicles get a warning message once they enter the 100 m range from the bad road. On the other hand, a road congestion alert to a vehicle remains valid for a certain time period. Therefore, we have content validity depending on application-specific requirements. This property will help to improve scalability in autonomous vehicles.

- *Content-centric networking*: Typically, vehicles are more interested in acquiring contents (such as traffic conditions) than in validating their original sources. Therefore, they receive contents from multiple other vehicles (without checking the source of the content), and process them to take improved decisions. Moreover, a content sent by a vehicle may not be the source of the content. The content may be forwarded by the vehicle, which received it from other vehicles. Content-centric networking will help vehicles to establish automatic operation and management. However, as mentioned earlier, adequate security mechanisms should be deployed so that no attacker can get access to the system.

- *Vehicular grid to vehicular cloud*: The vehicles are equipped with multiple sensors and actuators to sense the environment; with all these equipment, they can take appropriate decisions. Multiple electric vehicles constitute a vehicular grid (as mentioned earlier), which exchange energy with the main power grid to balance real-time energy supply and demand from customers. Electric vehicles can also constitute a vehicular cloud through the Internet service facility, in which, they can share real-time information about road conditions, and energy status of the power grid. Consequently, it will help vehicles to take coordinated decisions through a common platform, the vehicular cloud.

Figure 12.7 depicts the basic difference between ID (IP) centric VANETs and information-centric network of vehicles.

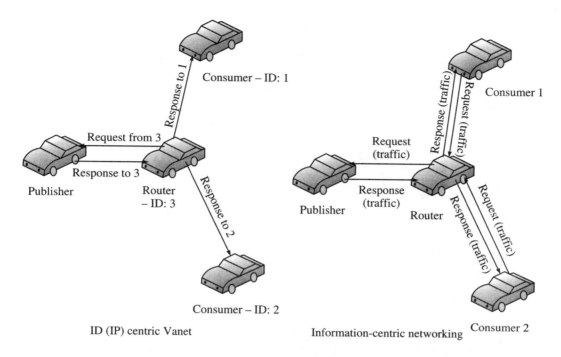

**Figure 12.7** Difference between VANET and information-centric networking

Another emerging approach is the integration of cloud services with the Internet of vehicles, in order to take better decisions. In such an approach, vehicles are facilitated with Internet services, as mentioned earlier. In addition to the information exchange between vehicles, there is also exchange of information with the cloud, which is further divided into roadside cloud (RSU) and central cloud. The vehicles exchange real-time information with RSU clouds once the former come within the clouds' vicinity. Further, RSU clouds aggregate information, and exchange the aggregated information with the central cloud. Therefore, vehicles on other roads also get the updated information. This principle is similar to that used in VANET. However, typically in a VANET, the vehicles are not integrated with on-board Internet service facility. Therefore, communication with the central cloud can only be done through the RSU clouds. In contrast, the integrated version of the Internet of vehicles and clouds enables both operations (traffic and data management), which, in turn, maximizes safety and provides improved decision making. Figure 12.8 presents a schematic view of the integrated architecture of the Internet of vehicles and clouds.

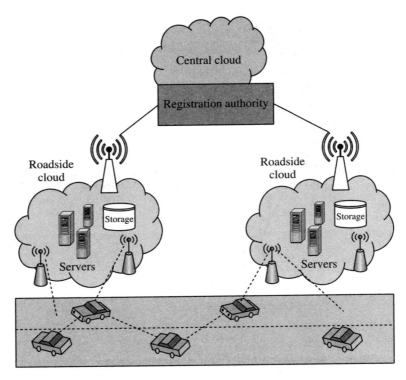

**Figure 12.8** Integrated architecture of the internet of vehicles and clouds

## 12.3 Future Trends and Issues

- We learnt about vehicle-to-grid energy exchange in smart grid in the presence of PHEVs. The discussed methods are mainly focused on a single-level energy exchange, i.e., vehicle to grid or grid to vehicle, which is a simple energy exchange model in the smart grid. We can have multiple levels of energy exchange in a V2G environment, for example, from a vehicle to other vehicles with larger capacity, and then to the micro-grids. Therefore, the vehicles with higher capacity can form a mobile energy grid, and supply energy to multiple micro-grids according to the real-time energy status. However, such energy exchange mechanisms require exhaustive study and experimented analysis prior to their real deployment in a smart grid environment.

- It is evident that PHEVs will play a vital role in both energy management and information exchange after it is integrated with the on-board Internet service facility. However, we have to ensure that the entire system is secure from any kind of attacks. For example, if any intruder injects any false information, the entire system may collapse. Additionally, as mentioned earlier, autonomous cars should have self-healing mechanisms to deal with adverse environment conditions.

- Finally, PHEVs should have the ability to charge their batteries from one micro-grid and supply that energy to other micro-grids that have energy deficit. Then, PHEVs can be used to regulate the energy service among multiple micro-grids. Therefore, we should have adequate mechanisms to exchange energy among multiple micro-grids. In the absence of adequate mechanisms, most of the excess energy in a micro-grid may be wasted, while other micro-grids may have energy deficit in the same time period.

## 12.4  Summary

In this chapter, we discussed different applications and the advantages of integrating plug-in hybrid electric vehicles (PHEVs) into smart grid. The concepts of vehicle to grid (V2G) and grid to vehicle (G2V) energy transfer were discussed. Then, different existing mechanisms that are useful to establish a smart environment in the smart grid were discussed. Finally, the chapter concluded by stating a few future trends and issues with PHEVs and smart grid.

### Test Your Understanding

Q01. Why are PHEVs attracting the attention of researchers in recent years?

Q02. Explain the concept of vehicle to grid and grid to vehicle.

Q03. State some advantages of integrating PHEVs into IoVs.

Q04. What is meant by optimal scheduling?

Q05. What are the three stages in the coalition game-theoretic approach?

Q06. What do you mean by inviters and invitees in the context of PHEVs?

Q07. Explain the concept of resource reservation.

Q08. State some emerging applications of vehicles facilitated with Internet of service.

Q09. What is the difference between VANET and the integrated version of the Internet of vehicles?

Q10. State some future trends and issues in plug-in hybrid electric vehicles (PHEVs).

## References

[1] Lin, J., K. C. Leung, and V. O. K. Li. 2014. 'Optimal Scheduling with Vehicle-to-Grid Regulation Service'. *IEEE Internet of Things Journal* 1 (6): 556–567.

[2] Yu, R., J. Ding, W. Zhong, Y. Liu, and S. Xie. 2014. 'PHEV Charging and Discharging Cooperation in V2G Networks: A Coalition Game Approach'. *IEEE Internet of Things Journal* 1 (6): 578–589.

[3] Kaur, K., A. Dua, A. Jindal, N. Kumar, M. Singh, and A. Vinel. 2015. 'A Novel Resource Reservation Scheme for Mobile PHEVs in V2G Environment Using Game Theoretical Approach'. *IEEE Transactions on Vehicular Technology* 64 (12): 5653–5666.

[4] Chu, P. C. and J. E. Beasley. 1998. 'A Genetic Algorithm for the Multidimensional Knapsack Problem. *Journal of Heuristics* 4 (1): 63–86.

[5] Yu, L., T. Jiang, and Y. Zou. 2016. 'Distributed Online Energy Management for Data Centers and Electric Vehicles in Smart Grid'. *IEEE Internet of Things Journal.* 3 (6): 1373–1384.

[6] Gutman, S. 1979. 'Uncertain Dynamical Systems–A Lyapunov Min-Max Approach'. *IEEE Transactions on Automatic Control* 24 (3): 437-443.

[7] 'Alternating Direction Method of Multipliers (ADMM)'. Accessed 05 November 2017. Available at: http://stanford.edu/boyd/admm.html.

[8] Gerla, M., E. K. Lee, G. Pau, and U. Lee. 2014. 'Internet of Vehicles: From Intelligent Grid to Autonomous Cars and Vehicular Clouds'. In *Proc. of the IEEE World Forum of Internet of Things (WF-IoT)*. pp. 241–246.

# CHAPTER 13

# Smart Buildings

Buildings are one of the most prominent energy consumers in modern-day power grids. In particular, commercial buildings consume large amounts of energy for operating lighting systems, heating–ventilation and air-conditioning (HVAC), and IT equipment including servers. High energy consumption at the buildings increases energy consumption cost and environment pollution. To deal with these issues, buildings can be converted to 'smart buildings'. Through the concept of smart buildings, electric loads can be controlled in an adaptive manner. Unused equipment can be switched off so that the total energy consumption in the building can be reduced. The appliances that are installed inside a building will generate huge data in order to have real-time information inside the building. To fulfill these objectives, different methodologies exist. In this chapter, we focus on commercial buildings as smart buildings for efficient energy management.

## 13.1 Concept of Smart Building

Most commercial buildings are equipped with HVAC systems, IT equipment, servers, and plug-in equipment. These equipments can be integrated with different sensing and actuation technologies, so that they can be controlled dynamically depending on their requirements. Different useful techniques that are deployed in different buildings to make them smart buildings, are discussed here.

- *Sensing*: It is the key technology required for setting up a smart building. Different sensors are installed in the building (inside and outside), so that different parameters (such as temperature, smoke, humidity, and motion) can be sensed in real-time. The sensed data are forwarded to the data center network for processing, computing, and decision making. For data collection and forwarding, wireless sensor network (WSN) is an emerging technology that can be used due to its unique features such as the capability of operating in low power. Typically, IEEE 802.15.4-based technologies (such as Zig-Bee and 6LowPAN) are used in WSN to forward real-time data. User context, environment, and energy consumption are three major factors to be monitored to establish a smart building. The sensors that are installed to monitor these components generate huge amount of data. This also requires suitable data management techniques for real-time energy management inside the building, as discussed in Chapter 11.

- *Actuation*: After sensing the users' contexts, environment, and energy consumption, actuation is another important task to be employed. HVACs should be controlled in an efficient manner considering the sensed information received from individual sensors. Whenever there is a significant change in the systems (monitored through the sensors), the actuators control (i.e., switches on/off) the equipment appropriately. For example, when there is no one in a room, lights can be switched off automatically to reduce energy consumption. Additionally, HVAC systems may be turned off when the temperature is moderate. Thus, the sensing and actuating systems control the building's energy consumption in an automated manner to reduce energy consumption, while considering the quality-of-service to its users.

Therefore, a building with automated energy management is termed as a *smart building*.

## 13.2 Challenges and Opportunities

There are several challenges and opportunities to transform a building to a smart building. Some of the challenges and opportunities are discussed here [1].

- *Sensor placement*: One of the most important challenges is sensor placement. Most modern buildings are integrated with smart sensors. How and where to place the sensors is a big issue to manage the real-time energy consumption in a building. Moreover, the global positioning system (GPS) does not work well inside the buildings. Therefore, suitable technologies are required to place the sensors optimally inside a building so that they can sense the environment, users' contexts, and energy consumption.

- *Components to be integrated*: As discussed before, different components consume energy inside a building. It may happen that some of the components cannot be powered off because of their properties and requirements. Therefore, a clear road-map is required to integrate different components in establishing a smart building.

- *Communication protocol*: Communication interference is a major challenge inside a building, as several components will communicate with the central entity. In the current day grid systems, existing communication protocols and technologies are used. This may not be efficient enough to fulfill the requirements of a smart grid. Therefore, suitable communication technology is also required so that the sensors can communicate with the central entity without increasing the interference problem.
- *Data management*: Data management is a challenging task for energy management inside a building. In contrast to the smart meter data management, a smart building generates heterogeneous data from various sources – appliances (such as light and air conditioners), and others (such as parking data and location data). Such multi-dimensional data need to be managed efficiently, in order to establish a smart building in smart grid.
- *Job scheduling*: Finally, job scheduling is another important task to be executed. A smart scheduler is required to schedule different appliances inside the smart building. For example, out of two different types of appliances, HVAC and lights, it is a challenge to determine which one should be prioritized during scheduling, while considering the requirements in real-time. It may happen that HVAC cannot be turned off at that moment, while the lighting system can be. All these requirements should be taken care of, before converting the building into a smart building.

## 13.3 Different Approaches for Establishing Smart Buildings

Several schemes are available to address the issues and challenges presented in Section 13.2. The existing schemes are discussed from two perspectives – energy management and information management.

### 13.3.1 Automatic energy management systems

#### Sensors and actuators-based management

As discussed earlier, sensing and actuation are the key enabling technologies to establish a smart building in the smart grid. The sensors are installed in individual appliances to monitor the real-time conditions and requirements. Based on the sensed information, appropriate decisions can be made in order to execute efficient energy management. After gathering the sensed information from individual sensors, it is important to have adequate control strategies. Considering this problem, Suryadevara et al. [2] proposed a WSN-based sensing and actuation technique for energy management in smart buildings. In such a system, the Zig-Bee protocol (based on IEEE 802.15.4) is used to communicate with the sensor devices. Figure 13.1 shows the schematic diagram of such a system integrated with sensors.

210 *Smart Grid Technology: A Cloud Computing and Data Management Approach*

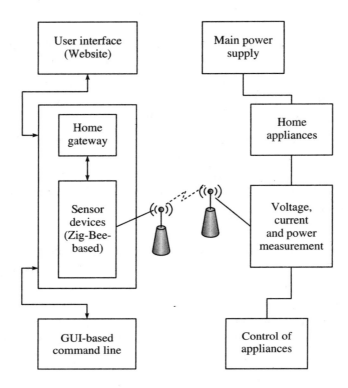

**Figure 13.1** Schematic view of WSN-based smart building system

As depicted in Figure 13.1, the user of the building can monitor the real-time energy consumption status through a web portal. The sensor nodes can be programed through GUI-based command line. The sensors measure different parameters such as voltage, current, and power at the appliances installed in the home. Based on this measurement, adequate control actions are taken to optimize the power usage by the appliances. The control of the appliances can be done from three different perspectives – automatic, remote, and manual control. In automatic control, suitable control logic is integrated with the sensing and actuating devices. On the other hand, in remote control, the authorized owner can access and control the appliances remotely through web access. In contrast to automatic and remote control, sensing and actuating devices can also be controlled manually, depending on real-time requirements. In this case, the user needs physical access to the devices. Based on the real-time situation, the aforementioned control strategies can be employed in order to have optimized electricity usage.

### Agent-based management

Modern day buildings are integrated with self-generated energy sources such as solar energy. Therefore, appliance scheduling is a promising approach to reduce the energy consumption from the main grid during peak hours, while consuming energy from the solar powered storage devices as much as possible. A minority game-based energy consumption scheme is presented in the presence of solar energy and main power grid [3]. In this scheme, the building consists of several rooms, where each room is equipped with smart meters. The smart meters report their energy requirements to the in-building energy management system (EMS). Based on the requested energy and expected outcome, the smart meters are scheduled to consume energy either from the storage device, or from the main grid, or both. Figure 13.2 depicts a system consisting of solar energy and main grid energy with different smart meters installed in different rooms inside a building.

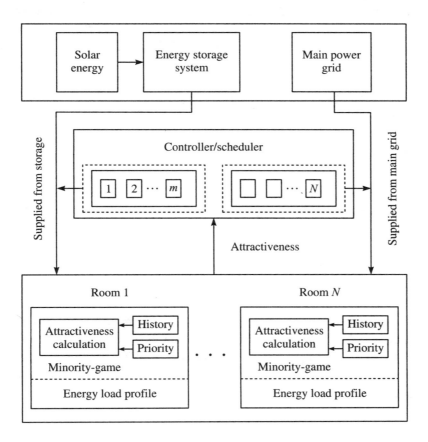

**Figure 13.2** Working flow of energy source scheduling in a smart building

In such a game-based scheduling approach, an attractiveness value is calculated to order the rooms in a rank-based manner. The following represent the mathematical calculation of the attractiveness value [3].

- *Preference factor calculation*: It denotes the preference for solar energy (over traditional energy) to be used in the future. Mathematically,

$$Pr_k(t) = \frac{E_k(t)}{\sum_{t=1}^{24} E_k(t)} \qquad (13.1)$$

where $Pr_k(t)$ denotes the preference factor at $t^{\text{th}}$ time period for room $k$ and $E_k(t)$ denotes the predicted energy demand at the time period $t$. It is to be noted that the total time period considered in this case is 24 hours. However, it can be decided based on the users' interests or application-specific requirements. Therefore, using the preference factor, a room calculates its willingness to consume energy from solar energy to reduce the energy consumption cost.

- *History factor calculation*: The history factor is used to balance the energy allocation, which is denoted mathematically as follows:

$$H_k(t) = 1 - \frac{S_k}{\sum_{k=1}^{N_r} S_k} \qquad (13.2)$$

where $S_k$ denotes the cost saving for room $k$ in the past, and $N_r$ is the total number of rooms participating in the energy trading process inside the building. Further, $S_k$ is represented as:

$$S_k = \sum_{\tau=0}^{t} \Delta p(\tau) \times E_{\text{solar}}^k(\tau) \qquad (13.3)$$

where $\Delta p(\tau)$ represents the price difference between the main power grid and the solar energy at time $\tau$. The price of solar energy can be conceptualized as the capital expenditure (CAPEX) and operational expenditure (OPEX) cost in energy production using solar panels. $E_{\text{solar}}^k(\tau)$ denotes the amount of energy allocated from solar energy at time $\tau$ for room $k$.

- *Attractiveness value calculation*: Finally, based on the preference and history factors, the value of attractiveness is calculated as follows:

$$A_k(t) = \alpha_k \times H_k(t) + (1 - \alpha_k) \times P_k(t) \qquad (13.4)$$

$\alpha_k$ is used to adjust the weight factors (preference and history) for different rooms. Thus, as presented in Figure 13.2, energy supply to the rooms is done using either solar energy or main grid energy according to the attractiveness values.

## Rule verification system

Different control logics are installed in the deployed devices inside a smart building system for flexible control and management of the system. Control logics are a set of service rules that are executed by the devices when taking any action. Consider a smart building system in which a large number of users are present. As a result, the number of required service rules to serve individual requirements of the users also increases, which, in turn, increases the complexity in the rule verification process. Therefore, an efficient rule verification system is required to be included within the smart building system. As mentioned by Sun et al. [4], existing rule verification systems focus on the conflict of logics between two rules, which may not be adequate to address the issues and challenges present in a smart building system. To meet the requirements of a rule verification system in a smart building system, a new framework is introduced [4]. Figure 13.3 presents a flow chart of the rule verification process proposed by Sun et al. [4].

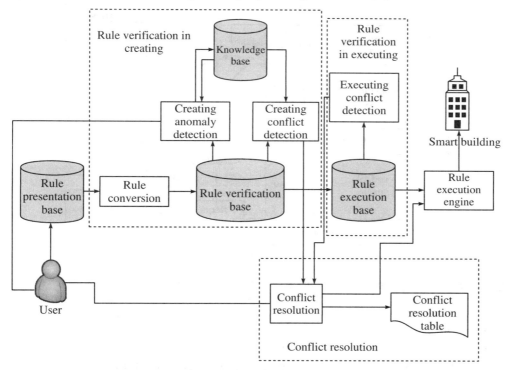

**Figure 13.3** Flow chart for rule verification process in a smart building

As the name suggests, *rule verification in creating* takes care of the verification of the rules while creating new ones. On the other hand, *rule verification in executing* verifies the rules while executing them. Finally, conflicts are addressed by the *conflict resolution* process. Detailed discussions on the aforementioned processes are not discussed as they are out of the scope of this book.

### Admission control inside the building

To control the appliances in an adaptive manner, it is required to implement an appliance controlling architecture inside a smart building. Recently, an architecture for smart appliance control was proposed [5], in which different interfaces are used to communicate with different entities. Figure 13.4 presents this architecture for smart appliance control inside the building in the smart grid system (adopted from [5]).

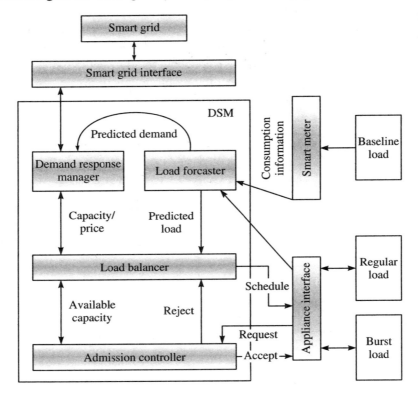

**Figure 13.4** Architecture for appliance control inside a smart building

- *Baseline load*: Baseline load is the total power consumption by appliances such as lighting, fan, and stove, which must be powered on in real-time. Therefore, during the admission control of other devices, the scheduler should take into account the total energy load from these appliances. These types of load may not be necessarily managed by the central unit. However, real-time energy consumption must be reported to the controller (such as smart meter) in order to calculate the total load from all types of appliances.

- *Burst load*: The appliances must be powered on for a particular time period, i.e., they have a fixed starting time and a finishing time. For example, a washing machine must be kept powered on for a fixed duration. Therefore, the peak load to the smart grid increases when such types of appliances are switched on. Consequently, a suitable scheduling technique is required to control the burst load, so that the peak load on the grid can be minimized.
- *Regular load*: Finally, regular loads can exist for daily appliances (such as refrigerator and heater). These loads can be interrupted even when appliances are operating. Therefore, the operation of such devices are controllable using admission control techniques. However, these types of appliances may also generate load bursts if they are switched on for a long time.

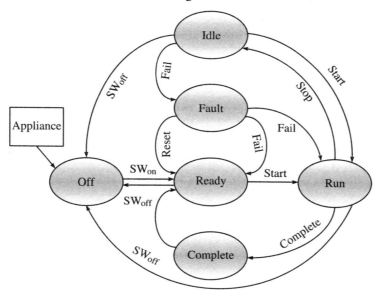

**Figure 13.5** Appliances finite state machine

Figure 13.5 presents the finite state transition diagram of the appliances installed inside a smart building (adopted from [5]). The controller takes into account all the transition steps before taking any decision.

### Job scheduling with uncertain local generation

Buildings can be facilitated with several local power generation sources such as solar and wind power. However, the intermittent behavior of such energy sources poses challenges to the power grid for efficient dispatch of electricity in real-time. Appliances inside the building can be scheduled appropriately to deal with such challenges. From this perspective, Danandeh et al. [6] proposed a job scheduling approach in the smart building system to deal with the uncertainty of locally generated energy supplies. The appliances are categorized according to their properties. Accordingly, some of the appliances can be scheduled ahead of their running time in which they will consume energy, whereas the rest of the appliances can be scheduled in real-time. Based on these two properties, the formulation a two-stage optimization problem is shown below. Mathematically, the optimization problem is formulated with a local generation $p_n^G$, as follows [6]:

$$\text{Minimize} \quad \sum_{n=0}^{N-1} \left( w_n^b C_n^b + w_n^e C_n^e - \rho_n C_n^b s_n \right)$$

subject to

$$\sum_{t=0}^{L_{ik}-1} \theta_{iktn} \leq y_{ik}, \forall i,k,n \tag{13.5}$$

$$\sum_{n=0}^{N-1} \theta_{iktn} = y_{ik}, \forall i,k,t \tag{13.6}$$

$$\sum_{n=0}^{N-1} \theta_{iktn} \geq \sum_{n=0}^{N-1} \theta_{ikt'n}, \forall i,k,t,t' \geq t \tag{13.7}$$

$$\theta_{iktn} \leq \sum_{n'=n+1}^{N-1} \theta_{ikt'n'}, \forall i,k,t,t' \geq t+1, n \tag{13.8}$$

$$\sum_{k=0}^{K-1} y_{ik} = 1, \forall i \tag{13.9}$$

$$\sum_{n=0}^{N-1} n \sum_{k=0}^{K-1} \theta_{ik0n} \geq A_i, \forall i \tag{13.10}$$

$$\sum_{n=0}^{N-1} n \sum_{k=0}^{K-1} \theta_{ik,L_{ik}-1,n} \leq B_i, \forall i \tag{13.11}$$

$$\sum_{n=0}^{N-1} n \sum_{k=0}^{K-1} \theta_{ik,L_{ik}-1,n} - \sum_{n=0}^{N-1} n \sum_{k=0}^{K-1} \theta_{ik0n} \leq R_i N + \sum_{k=0}^{K-1} (L_{ik}-1) y_{ik}, \forall i \tag{13.12}$$

$$\sum_{n=0}^{N-1} n \sum_{k=0}^{K-1} \theta_{ik,L_{ik}-1,n} + 1 \leq \sum_{n=0}^{N-1} n \sum_{k=0}^{K-1} \theta_{ik0n}, \forall i, j \in Pr_i \tag{13.13}$$

$$\sum_{i=0}^{I-1} \sum_{k=0}^{K-1} \sum_{t=0}^{L_{ik}-1} d_{ikt} \theta_{iktn} \leq P_n^{\max} + w_n^e + p_n^G, \forall n \tag{13.14}$$

$$w_n^b - s_n \geq \sum_{i=0}^{I-1} \sum_{k=0}^{K-1} \sum_{t=0}^{L_{ik}-1} d_{ikt}\theta_{iktn} - p_n^G - w_n^e, \forall n \tag{13.15}$$

$$w_n^b, w_n^e, s_n \geq 0, \theta_{iktn}, y_{ik} \in \{0,1\} \tag{13.16}$$

Therefore, the objective of the system is to minimize the energy consumption cost from the grid, while considering the profit in selling the energy back to the grid. Equation 13.5 denotes that only one segment of a job can be scheduled in one time period. Further, Equation 13.6 ensures that all the segments of a job are scheduled only once. Equation 13.7 confirms that each job is scheduled in the right order. A new segment can be served if and only if the service to the previous segment of the job is finished, which is ensured in Equation 13.8. Equation 13.9 denotes that only one mode of a job is selected, i.e., either the job must be scheduled in ahead of their running time or it can be scheduled in real-time. Equations 13.10 and 13.11 denote the time window of each job. Equation 13.12 denotes that a job cannot be interrupted during its execution time. Existing relationships between jobs are presented by Equation 13.13. On the other hand, Equations 13.14 and 13.15 denote the energy balance and the impact of step-wise price in the energy consumption, respectively. Finally, Equation 13.16 denotes that the values are positive; it also represents their boundary conditions in the optimization problem.

The aforementioned optimization problem can be solved using a two-stage linear programming approach [6]. In the first stage, all the appliances that must be scheduled in ahead of their running time can be scheduled considering all the constraints presented in the optimization problem. In the second stage, the appliances can be scheduled in real-time, while following the constraints presented in the optimization problem.

### 13.3.2 Intelligent information management systems

#### Machine to machine data collection

Machine to machine (M2M) communication is an important technology used in enabling smart grid. The corresponding data collection should be understood. Smart buildings can cooperate among themselves to forward the energy consumption data to the utility provider. From this perspective, a cooperative approach is proposed for smart grid data collection based on the M2M framework [7]. In such a setting, smart buildings form clusters, depending on their individual utility value, and forward the information in a cooperative manner. The cluster heads in the clusters can be static or dynamic, depending on the real-time scenario. Further, to form cluster heads and their members, a coalition game is employed, in which the smart meters take decisions on joining or leaving a cluster. Figure 13.6 presents a pictorial view of the clusters formed inside the smart building. Detailed discussion on the coalition game-theoretic approach is out of the scope of this book.

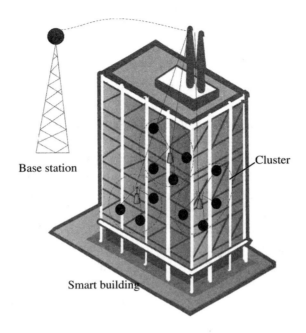

**Figure 13.6** Cluster formation inside a smart building

**Information exchange in the presence of smart buildings**
Smart buildings are an integral part of a smart grid. We have discussed earlier that a smart building consumes energy and sells energy back to the grid as well. Moreover, different buildings have different energy consumption profiles and economics, which are also dynamic in nature. Therefore, it is required to provide adequate real-time energy consumption information to both the grid and the building manager, in order to establish an economic energy exchange mechanism. Towards this objective, an information exchange framework is available for use in smart grids having smart buildings [8]. Typically, a micro-grid consists of several energy sources, such as (1) distributed local generations, which is uncontrollable (e.g., solar energy); (2) distributed generation, which is controllable (e.g., combined heat power); (3) storage devices (e.g., battery). A building also consists of different energy generation units such as solar and wind. Therefore, integration of smart buildings with micro-grid operation will facilitate the establishment of a balanced energy management scheme in the smart grid. For example, a building can buy energy from the micro-grid when there is a requirement of energy that cannot be served using the self-generated energy sources. On the other hand, a building can sell energy back to the micro-grid when there is a surplus of energy generated in the building. However, for efficient operation of energy exchange, we need a suitable information exchange mechanism, so that both the entities have real-time information of energy

consumption and the associated price. Based on the exchange of information, the optimization problem for a micro-grid can be formulated as follows [8]:

$$\text{Minimize} \quad \sum_{t=1}^{T} \left[ \hat{\lambda}(t) P_{\text{grid}}(t) + \sum_{i=1}^{N} \sum_{j=1}^{n_i} c_{ij}^t P_{ij}(t) \right]$$

subject to

$$\sum_{i=1}^{N} P_i(t) + P_{grid}(t) = \hat{P}_D^t, \forall t \tag{13.17}$$

$$P_{\text{grid,min}} \leq P_{\text{grid}}(t) \leq P_{\text{grid,max}}, \forall t \tag{13.18}$$

$$P_i(t) = P_{i,\min}^t + \sum_{j=1}^{n_i} P_{ij}(t), \forall i, \forall t \tag{13.19}$$

$$P_{ij,\min} \leq P_{ij}(t) \leq P_{ij,\max}, \forall i, \forall j, \forall t \tag{13.20}$$

$$P_k(t) = P_{k,\min}^t + \sum_{j=1}^{n_k} P_{kj}(t), \forall k, \forall j, \forall t \tag{13.21}$$

$$P_{kj,\min} \leq P_{kj}(t) \leq P_{kj,\max}, \forall k, \forall j, \forall t \tag{13.22}$$

$$E_i(t+1) = f_i(E_i(t), \Delta t P_i(t), \theta_i^t), \text{ for } i \in \mathcal{S} \cup \mathcal{F}, \forall t \tag{13.23}$$

$$E_{i,\min} \leq E_i(t) \leq E_{i,\max}, \text{ for } i \in \mathcal{S} \cup \mathcal{F}, \forall t \tag{13.24}$$

$$U_i(t) P_{i,\min}^+ \leq P_i^+(t) \leq U_i(t) P_{i,\max}^+, \text{ for } i \in \mathcal{S}, \forall t \tag{13.25}$$

$$(1 - U_i(t)) P_{i,\min}^- \leq P_i^-(t) \leq (1 - U_i(t)) P_{i,\max}^-, \text{ for } i \in \mathcal{S}, \forall t \tag{13.26}$$

$$P_i^+(t) = \sum_{j \in j^+} P_{ij}(t), P_i^-(t) = \sum_{j \in j^-} P_{ij}(t), P_i(t) = P_i^+(t) + P_i^-(t), \text{ for } i \in \mathcal{S}, \forall t \tag{13.27}$$

$$U_i(t) : \text{binary variable for } i \in \mathcal{S}, \forall t \tag{13.28}$$

where $P_{grid}$, $P_{ij}$, $P_i$, $E_i$, $P_i^+$, $P_i^-$, and $U_i$ are the decision variables, in which $i \in \mathcal{S}$. $\mathcal{S}$ denotes the set of storage devices and $\mathcal{F}$ denotes the set of flexible loads. Equations 13.23 and 13.24 denote the dynamics of the storage units and the flexible loads. On the other hand, Equations 13.25–13.28 denote the relationship between the segmented power level and the charging/discharging state. Therefore, the micro-grid and buildings manage the energy selling and buying process, while exchanging real-time information. Depending on the real-time price signal, micro-grid and buildings schedule their energy consumption profile dynamically.

### Control and communication protocols for load control

We have seen that a bi-directional communication network is the main backbone of the smart grid, in order to have real-time information on energy supply–demand and price at both the service providers' and customers' ends. Recently, control and communication protocols based on direct load control were proposed for smart building energy management [9]. In this approach, the energy is supplied to the appliances in the form of packets, and not the way it is done in traditional energy distribution. An energy packet can be considered as the required power to run a particular appliance for a certain duration. The length of the energy packet is variable over time and appliances. This is similar to the traditional packet switched networks. Therefore, the continuous flow of energy is packetized, as depicted in Figure 13.7.

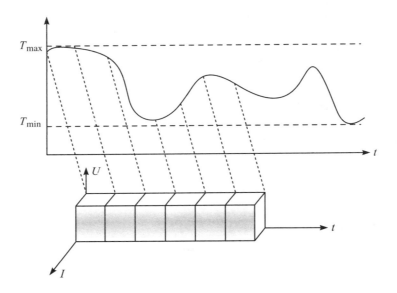

**Figure 13.7** Packetized view of energy supply in the smart grid

Consequently, two different communication scenarios are possible – full information and binary information communication – which are discussed here:

- *Full information communication*: Let us assume that energy is to be supplied for use by air-conditioning appliances. Therefore, at time $t$, a paired information $< T^i(t), 0(1) >$ consisting of desired temperature and operating status is sent to the smart building operator (SBO), where $0(1)$ denotes the operating status of the appliance, i.e., 0 denotes OFF and 1 denotes ON. The SBO reserves the required energy which will be supplied in a packetized form. Based on the real-time situation, this energy demand can be deviated to another time period, while considering the customer's comfort level. Therefore, based on the collected information from all appliances, the SBO dynamically predicts the evolution of $T(t)$ as [10]:

$$\frac{d}{dt}T(t) = \frac{T_{\text{out}} - T(t) - T_g u + w(t)}{\tau} \tag{13.29}$$

where $T_{\text{out}}$ is the outside temperature, $T_g$ the temperature gain when the appliance operates, $\tau$ the effective thermal time constant, $w(t)$ the bounded measurement error, and $u$ is the binary variable used to denote the status of the appliance. Finally, the minimum number of required energy packets considering all customers' requirements can be obtained as [11]:

$$m = \frac{N T_{\text{out}} - \sum_{i=1}^{N} T_{\text{set}}^i}{T_g} \tag{13.30}$$

- *Binary information communication*: After deciding the $m$ number of packets required to serve the energy demands from all customers inside the building, the SBO acquires real-time full information to deliver the energy packets. However, it may happen that due to the customers' privacy concerns and limitations in the communication system, full information is not available to the SBO. In such a scenario, the SBO adopts the binary information communication mechanism in which energy is served following a first-in-first-serve approach in real-time. After serving the $m$ number of energy packets, the SBO holds the next requests and serves their requests in subsequent intervals.

### Integrated smart home management

A smart building consists of several sensors that monitor different parameters inside the building. Therefore, the sensed data from the sensors is also required to be processed to take appropriate decisions in energy management. Towards this objective, a cloud-based smart home management framework is available while integrating community services in the smart grid [12].

As depicted in Figure 13.8, smart homes are controlled by multiple communities, and the communities are controlled by the cloud platform in a hierarchical manner. Firstly, the sensed information is processed at the home controller system. Then, the aggregated information is sent to the community service system, where the information collected from multiple smart homes are aggregated and processed. Finally, the aggregated information at the community broker are sent to the cloud platform, where final processing and computation takes place.

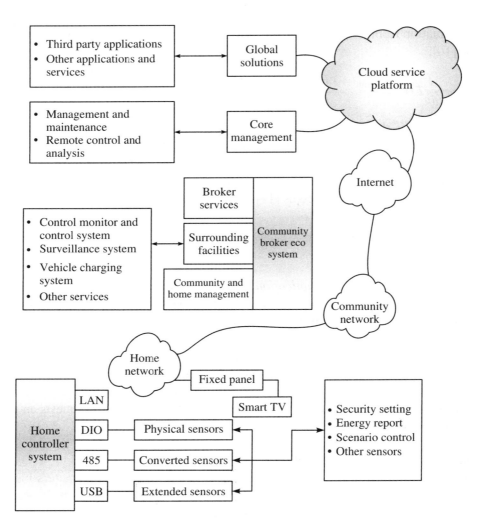

**Figure 13.8** Cloud-based smart home management

## 13.4 Future Trends and Issues

- In a WSN-based appliance control system, multiple communication technologies may co-exist. Therefore, the challenges involved in integrating multiple communication technologies in a single platform should be addressed from the perspective of fulfilling the smart grid requirements. Additionally, based on the received day-ahead energy usage and associated cost patterns, dynamic appliance scheduling schemes can be instituted in such a WSN-based energy management system.

- Renewable (non-dispatchable) energy sources are intermittent in nature; they also have high energy fluctuation. Therefore, the intermittent behavior of non-dispatchable energy sources needs to be taken care while integrating them into the building energy management systems. It is to be noted that several uncertainty-based energy trading schemes are proposed in the smart grid to address the problem of intermittent behavior of renewable energy sources. The existing schemes may require to be revisited to integrate them with smart building energy management systems.

- The existing rule verification system for smart building energy management in the smart grid is static or is manually configurable, as identified by Sun et al. [4]. Therefore, dynamic updation of the rules can be implemented while considering the dynamic knowledge about the system.

- The complexity of the job scheduling problem increases with an increase in the number of appliances. Therefore, for a building with large number of appliances, it may be difficult to schedule all the appliances in an optimized manner. Therefore, the existing solution approaches can be revised to support large-scale job scheduling inside the smart building.

- In addition to the direct load control based on energy packets, price-based energy packets can be introduced for smart building energy management. In such a case, the customers can request different sizes of energy packets, depending on the price of energy. For example, the size of energy packet is small while the real-time price is high, and vice-versa.

## 13.5 Summary

In this chapter, the concept of smart building was discussed. The different challenges and opportunities in establishing a smart building in the smart grid environment was also discussed. The existing approaches for establishing a smart building were discussed from two perspectives – energy management and information management. Finally, some of the future trends and issues were also presented.

### Test Your Understanding

Q01. What do you mean by a smart building?

Q02. Explain the techniques deployed in different buildings to convert them to smart buildings.

Q03. State some of the challenges and opportunities to make a smart building.

Q04. Describe sensor and actuator-based energy management in smart buildings.

Q05. Explain agent-based energy management in smart grid.

Q06. What is meant by baseline load, bursty load, and regular load?

Q07. State the different energy sources of a micro-grid.

Q08. Mention some future trends and issues in deploying smart buildings.

# References

[1] Samad, T., E. Koch, and P. Stluka. 2016. 'Automated Demand Response for Smart Buildings and Microgrids: The State of the Practice and Research Challenges'. In *Proceedings of the IEEE* 104 (4): 726–744.

[2] Suryadevara, N. K., S. C. Mukhopadhyay, S. D. T. Kelly, and S. P. S. Gill. 2015. 'WSN-Based Smart Sensors and Actuator for Power Management in Intelligent Buildings'. *IEEE/ASME Transactions on Mechatronics* 20 (2): 564–571.

[3] Huang, H., Y. Cai, H. Xu, and H. Yu. 2016. 'A Multi-agent Minority-game Based Demand-response Management of Smart Buildings towards Peak Load Reduction'. *IEEE Transactions on Computer-Aided Design of Integrated Circuits and Systems.* 36 (4): 573–585.

[4] Sun, Y., T.Y. Wu, X, Li, and M. Guizani. 2016. 'A Rule Verification System for Smart Buildings'. *IEEE Transactions on Emerging Topics in Computing.* 5 (3): 367–379.

[5] Costanzo, G. T., G. Zhu, M. F. Anjos, and G. Savard. 2012. 'A System Architecture for Autonomous Demand Side Load Management in Smart Buildings'. *IEEE Transactions on Smart Grid* 3 (4): 2157–2165.

[6] Danandeh, A., L. Zhao, and B. Zeng. 2014. 'Job Scheduling With Uncertain Local Generation in Smart Buildings: Two-Stage Robust Approach'. *IEEE Transactions on Smart Grid* 5 (5): 2273–2282.

[7] Luan, X., Z. Zheng, T. Wang, J. Wu, and H. Xiang. 2015. 'Hybrid Cooperation for Machine-to-Machine Data Collection in Hierarchical Smart Building Networks'. *IET Communications* 9 (3): 421–428.

[8] Joo, J. Y. and M. D. Ilic. 2016. 'An Information Exchange Framework Utilizing Smart Buildings for Efficient Microgrid Operation'. *Proceedings of the IEEE* 104 (4): 858–864.

[9] Zhang, B. and J. Baillieul. 2016. 'Control and Communication Protocols Based on Packetized Direct Load Control in Smart Building Microgrids'. In *Proceedings of the IEEE* 104 (4): 837–857.

[10] Ihara, S. and F. C. Schweppe. 1981. 'Physically Based Modeling of Cold Load Pickup'. *IEEE Transactions on Power Apparatus and Systems* PAS-100 (9): 4142–4150.

[11] Zhang, B. and J. Baillieul. 2012. 'A Packetized Direct Load Control Mechanism for Demand Side Management'. In *Proc. of the IEEE Conference on Decision and Control.* pp. 3658–3665.

[12] Lee, Y. T., W. H. Hsiao, C. M. Huang, and S-C.T. Chou. 2016. 'An Integrated Cloud-Based Smart Home Management System with Community Hierarchy'. *IEEE Transactions on Consumer Electronics* 62 (1): 1–9.

# Part IV
# Smart Grid Design and Deployment

# CHAPTER 14

# Simulation Tools

In Chapters 5–13, we discussed several methodologies that are useful to address different problems present in the smart grid in order to provide electricity to the users in a cost-effective and reliable manner. However, we also need to realize the system behavior and impact of the parameters used in a particular model through different controllable experiments. Thus, the use of suitable simulation tools is an important concern.

As we saw earlier, smart grid technologically combines two different entities–power grid and communication network. Therefore, a smart grid simulator needs to take into account the properties of both the power grid and the communication network. There are many tools available for the simulation of smart grid. In this chapter, we discuss some of the useful simulation tools that can be used for conducting smart grid experiments.

## 14.1 Simulation Tools

### 14.1.1 Open DSS

Open distribution system simulator (OpenDSS) [1] is one of the popular power grid simulation tools available. It deals with power grid system planning and analysis. As the name suggests, using this tool, one can model the power distribution system and analyze the behavior of the system prior to actual deployment. Primarily, it can help to analyze the needs of distributed generation units in a power grid system. Holistically, it supports future needs such as the modeling of smart grid applications, power delivery, and harmonics analysis. Some of the wide applications of OpenDSS simulation tools are as follows [1]:

- Power distribution planning and analysis
- Analysis of integrating multiple distributed generations
- Distributed storage modeling
- Wind power simulation
- Solar photo-voltaic simulations
- Impact of electric vehicles modeling and simulation
- Daily/annual energy load and generation simulation

By default, this simulator does not include any provision to have communication protocols that can be used to design the smart grid communication networks. However, it provides the flexibility of integrating communication models into it. It can also easily be integrated with MATLAB or other well-known simulators.

## 14.1.2 MATPOWER

MATPOWER [2] is another simulation tool designed for analyzing power problems in a power grid. This is a package that can be integrated with MATLAB. It is useful to simulate optimal power flow problems present in the smart grid. However, the features of simulation supported by this tool are purely related to power grid simulation. MATPOWER does not have any provision to model the communication networks.

## 14.1.3 NS-2 and NS-3

The most widely used network simulators are NS-2 (www.isi.edu/nsnam/ns/) and NS-3 (www.nsnam.org). Smart grid communication network can be designed and analyzed using these simulators. NS-2 and NS-3 provide the flexibility of designing different communication protocols that can be used in the TCP/IP model. Therefore, researchers working in the field of smart grid communication network, specifically, smart grid data aggregation, may use NS-2 or NS-3 to design any protocol for data aggregation in the smart grid. Eventually, the protocol can be integrated atop the network access layer. Similarly, we can have routing and transport layer protocols specifically designed for the smart grid communication network. However, these simulators do not simulate any provision to simulate the power grid. However, they can be integrated with MATLAB to design the power grid related factors. Thus, we can have simulation of both the communication network and the power grid.

### 14.1.4 GridSim

GridSim (http://www.buyya.com/gridsim) is primarily designed to simulate distributed systems having multiple entities such as users, brokers, and resources. It supports heterogeneous resources that can be integrated into a single platform using resource brokers. We can model complex distributed algorithms using this simulator. Some of the important applications/features that can be simulated using GridSim are as follows:

- Integration of heterogeneous resources and system behavior modeling
- Incorporation of an auction model, which is one of the important game-theoretic models used in distributed systems.
- Providing support for a communication network
- Providing support for the integration of real-traces to realize the real-system behavior

Based on the aforementioned features, GridSim can also be used to simulate smart grid experiments, while supporting both the distributed nature of the power distribution grid and the communication network.

### 14.1.5 OMNeT++

OMNeT++ (https://omnetpp.org/) is a simulation framework used in various domains. As reported by the developer, OMNeT++ itself is not a simulator; rather, it provides infrastructure and tools for simulating any system. Therefore, the existing models can be integrated into it to study the behavior of the integrated system. Some of the release areas in which OMNeT++ is used are as follows:

- Modeling of wired and wireless networks
- Simulation of any discrete event-based system
- Protocol modeling and performance analysis
- Hardware architecture modeling and analysis
- Performance evaluation of any complex system

Consequently, it can be used in simulating smart grid communication network with limited support of power grid simulation.

### 14.1.6 GridLAB-D

GridLAB-D (www.gridlabd.org/) is another important simulation tool used for modeling power grid systems. It captures all the latest technologies used in power delivery system. GridLAB-D can be used for the following purposes:

- Design of distribution automation system
- Peak load management – an important aspect of smart grid

- Modeling distributed energy generation and storage
- Dynamic rate structure analysis of electricity

Therefore, GridLAB-D can be used to model most of the smart grid energy systems with limited support of communication network.

### 14.1.7 SUMO

We observed in Chapter 12 that PHEVs play an important role in the smart grid energy management system. Therefore, simulation of PHEVs' mobility in an urban scenario is also important. SUMO (sumo.dlr.de/) is a simulator in which the mobility of PHEVs can be modeled for a given scenario. For example, if there are no real traces of mobility of PHEVs, SUMO can be used to generate mobility traces for a given city. The trace can be used to model the PHEVs' movement pattern in the city. It is to be noted that SUMO is used to generate traces, but not for simulating a system. It can be integrated with NS-3 or any other suitable simulator to model and analyze the system behavior.

## 14.2 Summary

In this chapter, we discussed the existing simulation tools that are useful to model and analyze the smart grid system, while highlighting the limitations. As can be observed, none of the simulation tools completely supports both the objective of a smart grid system–communication network and power grid network. A trade-off always exists, i.e., people working in the field of smart grid communication network modeling can prefer network simulators over power grid simulators. On the other hand, people working in the field of power grid network modeling can prefer power grid simulators over network simulators.

## References

[1] The Open Distribution System Simulator (OpenDSS), Electric Power Research Institute. Accessed 08 August 2017. Available at smartgrid.epri.com/simulationtool.aspx.

[2] A MATLAB Power System Simulation Package. Accessed 08 August 2017. Available at pserce.cornul.edu/matpovd/.

# CHAPTER 15

# Worldwide Initiatives

In Chapters 1–14, several concepts and associated simulation platforms were discussed from different theoretical perspectives. In this chapter, we present an overview of the smart grid initiatives taken by different agencies and countries, including the practical deployment of the different schemes, which were discussed earlier. The chapter is organized as follows – initiatives taken by large agencies (such as the European Union (EU)), ongoing and developed projects worldwide, and worldwide implementation of smart grid visions and roadmaps. At the outset, let us recall the objectives of the smart grid in brief:

- The use of bi-directional communication facility atop the existing power grid system in order to have real-time supply–demand information.
- The integration of self-generated energy sources into the existing power grid in order to relieve the peak load on the grid and to reduce the carbon footprint.
- Dynamic energy management of the smart grid, while enabling full cyber-security features.
- Customers' participation in the energy market.
- Supply–demand forecasting and prediction to have reliable and cost-effective energy management in the power grid.

To fulfill the objectives of the smart grid and its vision, several initiatives have been taken to realize the benefits of the smart grid system over the traditional power grid. Some of the initiatives are discussed in the next few sections.

## 15.1 Initiatives Taken by EU

The European Union is a forerunner in taking major smart grid initiatives. The main objective of the EU in introducing smart grids is customers' involvement in the energy trading. Primarily, the customers are considered as residential consumers, electric vehicles, and distributed generators. Establishing a connection between the grid and the customers for taking coordinated decisions is the primary objective of the EU. Distributed demand response, economic energy dispatch, and engagement of VPP are some of the key features of the EU smart grid initiatives. Table 15.1 presents a brief overview of the smart grid deployment status in Europe [1].

**Table 15.1** *Smart grid deployment status in Europe*

| Forecasted Investment | Funding for Development | Smart Meters Deployed or In-progress |
|---|---|---|
| EUR 56 billion by 2020 (estimated value) | • EUR 184 million–European funding under two projects.<br>• ≈ EUR 200 million from European Recovery Fund. | • More than 40 million smart meters are already installed.<br>• 240 million will be installed by 2020. |

The European Technology Platform (ETP) smart grid was established in 2005 to start the smart grid initiatives in Europe and beyond. The platform consists of several representatives from R & D, industry, academia, and practitioners. Figure 15.1 presents a schematic overview of the smart grid network proposed by ETP (adopted from [2]).

As shown in Figure 15.1, the smart grid consists of a traditional centralized power plant. Further, CHP, renewable energy sources, and fuel cells, which are distributed in nature, are combined together with the traditional power grid system. Moreover, all these distributed energy sources can also form a VPP to provide uninterrupted energy supply to the customers.

The EU classified smart grid projects according to their application areas as follows:

- Smart network management – focuses on the bi-directional communication in the smart grid and other communications aspects as well.

- Integration of distributed energy resources – takes care of integrating distributed energy sources, such as wind and solar energy into the main power grid. The energy sources are integrated in a small-scale basis, for example, inside a smart building.

- Integration of large-scale renewable energy sources – focuses on the large-scale setup of renewable energy sources, while taking care of issues in carbon footprint and green energy requirement as well.

- VPP – connects multiple energy resource units into a common platform in order to provide uninterrupted energy supply to the customers. This also helps in abstracting the energy generation units from the customers.
- Smart customers and homes – these types of projects focus on the establishment of smart homes. Additionally, that also focus on the customers' engagement in real-time energy trading by implementing different policies.
- PHEVs – focus on PHEVs management and pricing policy in V2G and G2V scenarios.
- Others – apart from the aforementioned areas, there can be other projects that focus on smart grid development as well.

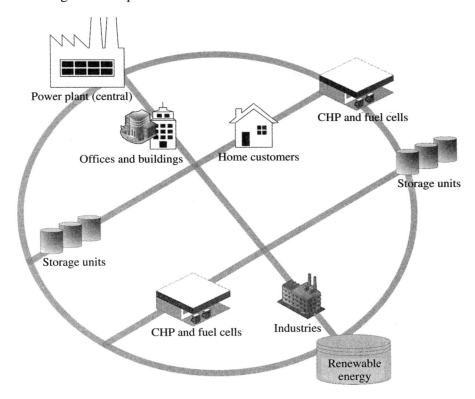

**Figure 15.1** Schematic overview of smart grid network according to ETP

It is forecasted that approximately 240 million smart meters will be deployed by 2020 in the EU [3]. Table 15.2 presents an overview of the distribution of projects in European countries [1].

**Table 15.2** *Distribution of projects in European countries*

| Country Name | Forecasted Status (%) |
|---|---|
| Austria | 6.1 |
| Belgium | 4.2 |
| Denmark | 22.0 |
| Finland | 1.5 |
| France | 4.2 |
| Germany | 11.1 |
| Italy | 5.5 |
| The Netherlands | 6.8 |
| Portugal | 2.4 |
| Spain | 8.7 |
| Sweden | 5.0 |
| the United Kingdom | 6.8 |
| Bugaria | 0.4 |
| Cyprus | 0.4 |
| Czech Republic | 1.7 |
| Estonia | 0.2 |
| Hungary | 1.1 |
| Latvia | 0.4 |
| Lithuania | 0.7 |
| Malta | 0.4 |
| Poland | 1.7 |
| Romania | 0.6 |
| Slovakia | 3.1 |
| Slovenia | 0.7 |

In terms of deployment, the EU categorized smart grid development into multiple levels as follows:

- *Level 0*: New energy generation techniques – first of all, there should be a focus on the new energy generation techniques available to meet the increasing energy demand of customers. The current energy generation units have limited capability to provide energy to the customers. However, new generation facilities to meet the increasing demand should be explored. Otherwise, there may be shortage although other technologies are in place.

- *Level 1*: Smart transmission network – once Level 0 is complete, the focus should be on the transmission network, to ensure that it is capable of transmitting huge amounts of electricity. The EU plans to obtain or create affordable technologies to have smart power transmission network all over Europe. The transmission network should also be capable of detecting huge loss or energy theft.

- *Level 2*: Smart communication network and processes – bi-directional communication facility is another important aspect in the smart grid. So, once Levels 0 and 1 are complete, smart communication network should be implemented on top of the transmission and distribution systems. Additionally, the information generated from such systems should be processed in real-time. Consequently, we need to have information technology (IT) analytics to enable such features in the smart grid.

- *Level 3*: Smart integration – this focuses on the integration of multiple energy sources into a single platform. The integration includes distributed energy sources, electric vehicles, and storage and aggregation. In addition to the energy integration, information aggregation is also an important factor. Therefore, in this level, the integration of energy sources and information aggregation are taken into consideration.

- *Level 4*: Smart energy management–once Levels 0 – 3 are complete, the underlying infrastructure of a smart grid system will be ready. With the help of generated units, distributed generations and information aggregation, real-time energy management will be done at this level. This includes distributed demand response and dynamic pricing.

- *Level 5*: Smart customers – finally, customers' participation in the smart grid energy management is a crucial factor. Therefore, intelligent policies should be deployed to engage the customers in real-time energy trading.

In order to implement smart metering infrastructure, EU introduced two interfaces–H1 and H2 [5].

- *H1*: It defines the interfaces between the smart meter and the external device. The communication between the smart meter and the external device in one-way. For example, a smart meter is connected to an external device to display the current status (such as energy supply, demand, and price) of the smart grid.

- *H2*: It defines the interfaces among devices, in which bi-directional communication is involved. For example, a smart meter is connected to a local gateway device to exchange real-time information. In such a scenario, the smart meter uploads its energy consumption information to the gateway. On the other hand, the gateway sends real-time energy price information to the smart meter. To be more specific, an interface is termed as an H2 interface if smart grid demand response is involved through that interface.

Figure 15.2 depicts the H1 and H2 interfaces defined by the Smart Grids Task Force Expert Group [5].

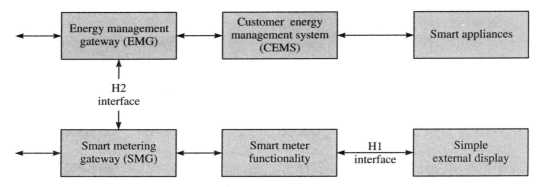

**Figure 15.2**  H1 and H2 interfaces used in smart grid communication

In order to address the interoperability issues present in smart grid, EU also defined a few steps to be considered in the process [5]. Figure 15.3 presents the main steps to be considered to address the issues of interoperability present in smart grid.

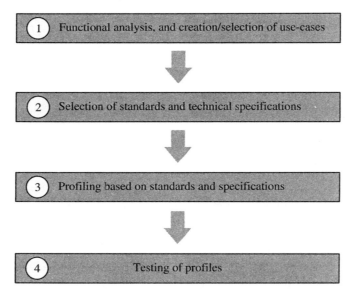

**Figure 15.3**  Steps to address interoperability issues in smart grid

## 15.2 Initiatives Taken by US Department of Energy

Concurrent to the EU, the US Department of Energy (DOE) also started initiatives on smart grid deployment in the US. Overall, the objectives of the US DOE behind smart grid deployment are similar to those of the EU. According to the US DOE, the smart grid initiative was taken into consideration to achieve the following goals:

- Use of digital communication over the existing power grid to have reliable and efficient energy management.
- Increased security features to deal with loss and energy theft.
- Deployment of smart devices, such as smart meters, to have real-time energy information from both ends – generation and distribution.
- Deployment of smart technologies to support the smart devices in the system.
- Integration of distributed generation units with the main power grid.
- Integration of electric vehicles into the energy trading system, in order to maintain real-time supply-demand balance.
- Active participation of customers' in the real-time energy market.

Table 15.3 presents the ongoing development of smart grid initiatives in the US [4].

**Table 15.3** *Smart grid deployment status in the US*

| Forecasted Investment | Funding for Development | Smart Meters Deployed or In-progress |
| --- | --- | --- |
| $338 to $476 billion by 2030 (estimated value) | $9.6 billion in 2009 under US Recovery Act. | • 8 million in 2011.<br>• 60 million by 2020. |

As in the EU, the US DOE also classified the smart grid project into the following categories:

- Advanced metering infrastructure – as discussed in Chapter 9, advanced metering infrastructure plays an important role in presenting real-time supply – demand information to utility providers. Therefore, DOE classified those projects under this category, which are useful for realizing advanced metering infrastructure. This also includes bi-directional information management in the smart grid system.
- Energy transmission systems – as discussed earlier, the transmission system also needs to be modified to support the increasing demand of customers. Projects related to the improvement of the energy transmission system falls into this category.
- Energy distribution systems – similar to the transmission system, there is another category which focuses on the energy distribution system. The issues related to the

distribution side are taken care of in this category. For example, micro-grid level management, in which the energy from the main grid is regulated to the customers in a distributed manner, while integrating renewable energy sources.

- Integrated systems – in this category, multiple energy generation units and systems are integrated in order to establish VPP and take coordinated decisions in providing cost-effective and uninterrupted energy supply to the customers.
- Storage systems – in smart grid, the storage units also play a crucial role in real-time energy management. The projects related to energy storage are classified into this.
- Smart homes or buildings – finally, the establishment of smart homes or buildings is another aspect, while implementing new equipment. Smart intelligent decision making approaches are also considered into this category.

Smart grid projects in the US are primarily funded through DOE, which includes the following [6]:

- Smart Grid Investment Grants (SGIG)–99 projects
- Smart Grid Demonstration Projects (SGDP)–32 projects
- Renewable and Distributed System Integration (RDSI)–9 projects

Table 15.4 shows the investment in smart grid projects in the classified areas according to DOE.

**Table 15.4** *Smart grid investment on grant projects*

| Project Category | Sharing Percentage (%) |
|---|---|
| Cross cutting projects | 39 |
| Advanced metering infrastructure | 30 |
| Electric transmission | 10 |
| Electric distribution | 13 |
| Customers' systems | 5 |
| Equipment manufacturing | 2 |

There are also initiatives for the standardization of systems in the smart grid.

## 15.3 Smart Grid Initiatives in Other Countries

Concurrent to the EU and DOE, a few other countries also initiated smart grid projects. This includes China and India, to name a few [7].

### 15.3.1 Initiatives in China

The main features of smart grid initiatives in China include:

- Defining policy and strategy for smart grid development.
- Training manpower and preparing engineering task force to analyze the impact and emergence of the smart grid system.
- Dealing with interoperability issues present in the system, while considering multiple systems to be combined together.
- Integration of renewable energy sources.
- Studying the impact of electric vehicles in the smart grid system.

Overall, the smart grid developments in China are categorized into three phases: (a) planning and testing; (b) construction and development; (c) upgradation of existing systems.

### 15.3.2 Initiatives in India

In India, there are few major issues and challenges, in addition to the aforementioned ones for EU and DOE. The major issues and challenges in the power grid system in India are as follows:

- Poor planning of distribution system – the entire distribution system needs to be modified in order to implement smart grid technologies. The existing system is poorly managed, which, in turn, makes the integration of new technologies ineffective.
- Energy theft at the distribution level – energy theft is another important issue in the energy market in India. A proper security and regulatory act should be introduced to deal with such issues.
- Overloading on the existing system – often, the existing systems are overloaded due to the poor management and maintenance of the distribution system. This problem leads to load-shedding in the distribution side, although the total energy supply is sufficient to fulfill the requirement of the total demand from customers.
- Manual billing system – finally, the major drawback in the energy management system in India is manual billing. Manual billing should be converted to automatic meter reading and billing of the customers.

Recently, the Government of India started smart grid initiatives, while considering the following measures:

- Separation of generation, transmission, and distribution units – earlier, a single provider was responsible for energy generation, and transmission and distribution. Presently, it is separated into multiple units, and each unit is taken care of by independent authorities for improved energy service.

- Implementation of digital meters and its efficiency in customers' billing and collection.
- Efficient use of electricity to reduce the overall energy consumption cost.
- Integration of renewable energy sources into the main power grid.
- Implementation of new generation units to cope with increasing demand from customers.

In addition to the challenges for AMI mentioned in Chapter 9, few other challenges for AMI in India include the following [9]:

- Lack of finances
- Limited skilled manpower
- Lack of smart grid awareness
- Lack of customer engagement
- Lack of universal standards

The Government of India also started a few pilot projects on smart grid in few states/cities, which are presented in tabular format [8], as shown in Table 15.3.

**Table 15.5** *Smart grid initiatives in India*

| State | Organization | Consultant | Amount (INR) |
|---|---|---|---|
| Assam | APDCL | Medhaj Techno Concept Pvt. Ltd. | 29.94 Cr |
| Haryana | UHBVN | n/a | 20.7 Cr |
| Himachal Pradesh | HPSEB | POWERGRID | 19.45 Cr |
| Mysore | CESC | POWERGRID | 32.59 Cr |
| Punjab | PSPCL | POWERGRID | 10.11 Cr |
| Telengana | TSSPDCL | CPRI | 41.82 Cr |
| Tripura | TSECL | POWERGRID | 63.43 Cr |
| West Bengal | WBSEDCL | POWERGRID | 7.03 Cr |
| Puducherry | PED | POWERGRID | 46.11 Cr |

Note: 1 Cr = USD 160,000 (approx.).

## 15.4 Smart Grid Standards

Several standards have been introduced to standardize smart grid technology, which includes standardization of architecture, communication technology, security, operating current and voltage, and many others. Table 15.6 summarizes the currently available standards for smart grid technology [10].

**Table 15.6** *Smart grid standards and their brief description*

| Standards | Description |
|---|---|
| Open Geospatial Consortium Geography Markup Language (GML) | This focuses on geographical locations' data requirements for smart grid applications. |
| IEC 62351 | It defines security requirements of a power system. |
| IEC 61851 | It defines the operating supply voltage for charging electric vehicles. The standardized value for charging an electric vehicle is 690 V (for AC) and 1000 V (for DC). |
| IEEE 1686-2007 | It defines the functions and features to be included in intelligent electronic devices (IEDs) installed at substations. The standard focuses on the security features that should be included in substations. Further, this includes access, firmware definitions, and data retrieval using IEDs at substations. |
| SAE J2836/1 | This standard defines parameters used for communication between electric vehicles and smart grid. The communication primarily corresponds to V2G and G2V applications. |
| IEEE 1815 (DNP3) | It focuses on the automation of substations, which further includes communication between substations and control centers. |
| IEEE C37.118-2005 | It defines specifications of the phasor measurement unit (PMU). Therefore, it focuses on the communications of synchrophasor data. |
| IEEE 1547 Suite | This standard defines the (physical) connections between the main power grid and distributed generations. |

| Standards | Description |
| --- | --- |
| IEEE 1588 | This standard is introduced to have synchronization among all components used in the smart grid. It helps in consistent time management in the smart grid. |
| Internet protocol for smart grid | Typically, IPv4/IPv6 is used for packet delivery in the smart grid. However, IPv6 is more common than IPv4 as large number of devices can be supported using the former. |
| IEEE 1901-2010 (ISP) | It defines broadband over power line networks. This standard focuses on the opportunities to provide Internet connection through electrical power lines. Using such a technology, information can be collected from smart meters without having a dedicated communication network. |
| SG-AMI 1-2009 | This focuses on the upcoming policies for upgradation of smart grid technology. It is used by smart meter suppliers, utility providers, stakeholders, and other parties related to smart grid energy management. |
| Open Automated Demand Response (OpenADR) | It defines specification of the message exchanged between customer and DR utility providers. |
| OPC-UA Industrial | This defines specification of an independent platform for secure, reliable, and fast message exchange, in order to support modern publish-subscribe (pub-sub) architecture. Therefore, it will help in establishing a service-oriented architecture (SOA) in the smart grid comprising utility providers, customers, and third parties. |

## 15.5 Summary

In this chapter, smart grid initiatives taken worldwide were discussed. The EU and DOE have made significant progress in deploying smart grid systems. China and India have also started deploying smart grid technologies in their existing power grid system. To summarize, smart grid is expected to become the dominant technology in the global energy market; it is what is needed in order to have reliable and cost-effective energy management.

# References

[1] Giordano, V. and S. Bossart. 2012. 'Assessing Smart Grid Benefits and Impacts: EU and U.S. Initiatives'. *Joint report of EC JRC and US DOE, JRC 73070*.

[2] European Commission. 2006. *European SmartGrids Technology Platform: Vision and Strategy for Europe's Electricity Networks of the Future*. Luxembourg: Office for Official Publications of the European Communities.

[3] Pike Research. 2011. 'Smart Grids in Europe, Pike Research Cleantech Market Intelligence'. Accessed February 2011. Available at http://www.pikeresearch.com/research/smartgrids-in-europe,

[4] Tech Report. 2011. *Smart Meters Deployment Looks Strong for 2011*.

[5] European Smart Grids Task Force Expert Group 1. 2016. 'Standards and Interoperability'. Available at https://ec.europa.eu/energy/sites/ener/files/documents/EG1_Final

[6] U.S. Department of Energy (DOE). 2009. '*Smart Grid System Report*'. DOE Report.

[7] Hashmi, M. 2011. *Survey of Smart Grids Concepts Worldwide*. Finland: VTT. Available at http://www.vtt.fi/publications/index.jsp.

[8] India Smart Grid Knowledge Portal. 'Smart Grid Pilot Projects in India'. Accessed October 2017. Available at http://www.indiasmartgrid.org/pilot.php.

[9] India Smart Grid Knowledge Portal. Accessed 05 November 2017. Available at http://www.indiasmartgrid.org/resourcecenter.php,

[10] India Smart Grid Knowledge Portal. Accessed 05 November 2017. Available at http://www.indiasmartgrid.org/standard.php.

# Index

Advanced metering infrastructure, 12, 118
Alternating direction method of multipliers, 201
ancillary demand response, 62, 66
appliance scheduling, 66
Application programming interface, 23
Approximation algorithm, 56
Asymmetric key cryptography, 149

Baseline load, 214
Big data, 38
Bilinear map group, 161
Bilinearity, 161
Billing policy, 82
Binary information communication, 221
Bloom filter, 138
Brute-force attack, 153
Building energy management, 8
Burst load, 215

Certificate authority, 151
Certificate revocation list, 134, 151
Certificate Revocation, 138
Chosen cipher text attack, 152

Chosen plain-text attack, 152
Cipher-text attack, 152
Cipher-text, 148
Clear-to-send, 121
Client-server architecture, 23
Cloud computing, 18
cloud-based demand response, 67
Cluster computing, 25
Clustering dispersion indicator, 173
Collision attack, 153
Collision resistant, 150
Combined heat power, 106
Commercial virtual power plant, 108
Community cloud, 34
Computability, 161
Cryptanalysis, 152
Cyber attack, 140
Cyber physical system, 147

Data aggregator unit, 97
Data attack, 140
Data storage, 118
Day-ahead pricing, 97
Decryption, 148

demand response, 61
Demand scheduling, 64
Demand side management, 8
Denial of service, 133
Dimension reduction method, 55
Direct clustering, 173
Discrete Fourier transform, 173
Dispatchable energy, 9
    source, 13
Distributed generation, 154
Dynamic pricing, 96

economic demand response, 62–63
Electro-geographical area, 109
emergency demand response, 62. 65
Encryption, 148
Energy generation, 234
Energy management system, 114
Energy theft detection, 140

First search and find of density peaks, 174
Frobenius norm, 178

Geography Markup Language, 241
Global positioning system, 208
Grid computing, 26
Grid to vehicle, 112, 189
GridLAB-D, 229
GridSim, 229

Hardware virtualization, 28
Heating–ventilation and
    air-conditioning, 207
hidden Markov model, 173
Home energy demand, 99
Home energy management, 8
    unit, 119
Hybrid cloud, 34
Hypervisor, 29

Identity-based encryption, 181

Indirect clustering, 173
Infrastructure as a service, 31
Integer linear programming, 50, 121
Internet of energy, 109
Internet-of-vehicle, 190
Islanded information, 172

K-L distance method, 175
K-L distance, 176
Key management function, 131
Known plain text attack, 152

Linear programming, 49
Load controller, 84
Load profiling, 172
Load-balancing, 82
Low-power lossy networks, 123
Lyapunov optimization, 201

Machine to machine, 217
Mainframes, 20
Man-in-the-middle attack, 152
Markle tree, 138
MATPOWER, 228
Mean index adequacy, 173
Message digest, 148
Micro-Grid, 13
mid-peak, 62
Mixed integer linear programming, 51

Non-degeneracy, 161
Non-dispatchable energy, 9
    source, 13
Non-linear programming, 51
Non-renewable energy, 9
Normal distribution, 53
NS-2, 228
NS-3, 228

off-peak, 62
OMNeT++, 229
on-peak, 62

Open DSS, 227
Outage management, 184

Partial usage patterns, 177
Pay-per-use model, 28
Physical attack, 139
Piece-wise aggregate approximation, 175
Pilot projects in India, 240
Platform as a service, 31
Plug-in hybrid electric vehicle, 14, 64, 189
Power Line Communication, 12
Pre-image resistant, 149
Predictive risk analysis, 184
Principal component regression, 55
Principle component analysis, 173
Private cloud, 33
Private key, 160
Projects distribution in the EU, 233
Public cloud, 33
Public key cryptography, 150, 160
Public key infrastructure, 149
Public key, 160

Quadratic cost function, 99
Quadratic function, 52
Queuing delay, 87

Reduced rank regression, 55
Regular load, 215
Renewable energy, 9
Replay attack, 153
Request-to-send, 121
Road side unit, 203
Roaming energy demand, 99

Scatter index, 173
Second preimage resistant, 149
Semi-structured data, 41
Service level agreement, 86, 115
Service oriented architecture, 27
SG deployment status in Europe, 232

shiftable, 61
Singular value decomposition, 177
Smart building operator, 220
Smart grid development status in the US, 237
Smart grid framework, 3
Smart grid in China, 239
Smart grid in India, 239
Smart grid initiatives, 231, 235
Smart grid interoperability, 236
Smart grid projects in the US, 238
Smart grid standards, 241
Smart Grid, 3
Smart integration, 235
Smart meter data management system, 122
Smart Meter, 12
Smart transmission network, 235
Software as a service, 32
State of charge, 193
Structured, 41
SUMO, 230
Support vector machine, 177
Symbolic aggregate approximation, 173
Symmetric key cryptography, 148

Technical virtual power plant, 108
The Gaussian distribution, 54
The Poisson distribution, 54
Transmission delay, 87

Unconstrained optimization, 52
Unstructured data, 41
Usage-based dynamic pricing, 98
Utility computing, 28

Vehicle to grid, 112, 189
Virtual integration, 114
Virtual power plant, 106
Virtualization, 28

Wireless sensor network, 208